ENCYCLOPAEDIA OF
STATISTICS

ENCYCLOPAEDIA OF STATISTICS

Surender Sachit

ANMOL PUBLICATIONS PVT. LTD.
NEW DELHI - 110 002 (INDIA)

ANMOL PUBLICATIONS PVT. LTD.
H.O.: 4374/4B, Ansari Road, Darya Ganj,
New Delhi-110 002 (India)
Ph.: 23278000, 23261597

B.O.: No. 1015, Ist Main Road, BSK IIIrd Stage
IIIrd Phase, IIIrd Block
Bangalore - 560 085 (India)
Visit us at: www.anmolpublications.com

Encyclopaedia of Statistics

First Published, 2006

ISBN 81-261-3031-8

PRINTED IN INDIA

Printed at Mehra Offset Press, Delhi.

Preface

The Encyclopaedia of Statistics covers all aspects of Statistics. This Encyclopaedia covers a range of various Statistics disciplines such as Mathematical Statistics and Probability, Applied Statistics and Data Analysis, Actuarial Statistics, Biostatistics and Data Analysis, Statistical Quality Control and Operations Research, Quantitative Economics and Planning ,Advanced Probability and Related Topics .

This new encyclopaedia integrates all of the diverse approaches that have gone into the making of the modern field of Statistics. It is unique among available reference works in that it covers the most recent topics of research. Encyclopaedia of Statistics is intended for all those interested in this field. It will be of immense value to students of all kinds, especially students of different fields of Statistics.

It will also be a major source of reference for professionals and students, who require an authoritative and up-to-date guide to assist them in their work. In preparation of this book, the author has freely consulted large number of books and journals so no authenticity is claimed. Author is especially thankful to Anmol Publications Pvt. Ltd., New Delhi for shaping this book in its final form. Suggestions for further improvement of this book are not only welcome but also greatly appreciated.

Author

A

Abrupt permanent impact In time series, A permanent abrupt impact pattern simply implies that the overall mean of the times series shifted after the intervention; the overall shift is denoted by ω (omega).

Abrupt temporary impact In Time Series, the abrupt temporary impact pattern implies an initial abrupt increase or decrease due to the intervention which then slowly decays, without permanently changing the mean of the series. This type of intervention can be summarized by the expressions:

Prior to intervention: $Impact_t = 0$

At time of intervention: $Impact_t = \omega$

After intervention: $Impact_t = \delta\ Impact_{t-1}$

Note that this impact pattern is again defined by the two parameters δ (delta) and ω (omega). As long as the δ parameter is greater than 0 and less than 1 (the bounds of system stability), the initial abrupt impact will gradually decay. If δ is near 0 (zero) than the decay will be very quick, and the impact will have entirely disappeared after only a few observations. If δ is close to 1 then the decay will be slow, and the intervention will affect the series over many observations. Note that, when evaluating a fitted model, it is again important that both parameters are statistically significant; otherwise one could reach paradoxical conclusions. For example, suppose the ω parameter is not statistically significant from 0 (zero) but the δ parameter is; this would mean that an intervention did not cause an initial abrupt change, which then showed significant decay.

Abs. E. mean Average absolute error (difference between target and actual output values) of the output variable.

Accept support (AS) testing In this type of statistical test, the statistical null hypothesis is the hypothesis which, if true, supports the experimenter's theoretical hypothesis. Consequently, in AS testing, the experimenter would prefer not to obtain "statistical significance."

In AS testing, accepting the null hypothesis supports the experimenter's theoretical hypothesis.

Activation function A function used to transform the activation level of a unit (neuron) into an output signal. Typically, activation functions have a "squashing" effect. Together with the PSP function (which is applied first) this defines the unit type.

Neural Networks supports a wide range of activation functions. Only a few of these are used by default; the others are available for customization.

Activation functions Function Definition Range

Identity X (-inf,+inf)

Logistic $\frac{1}{1-e^{-1}}$ (0,+1)

Hyperbolic $\frac{e^{x}-e^{-x}}{e^{x}+e^{-x}}$ (-1,+1)

-Exponential e^{-x} (0, +inf)

Softmax $\frac{e^{x}}{\sum_i e^{x_i}}$ (0,+1)

Unit sum $\frac{x}{\sum_i x_i}$ (0,+1)

Square root $\sqrt{x}$ (0, +inf)

Sine sin(x) [0,+1]

Ramp $\begin{Bmatrix} -1 & x \le -1 \\ x & -1 < x < +1 \\ +1 & x \ge +1 \end{Bmatrix}$ [-1,+1]

Step $\begin{Bmatrix} 0 & x < 0 \\ +1 & x \ge 0 \end{Bmatrix}$ [0,+1].

Addition rule The addition rule is a result used to determine the probability that event A or event B occurs or both occur.

The result is often written as follows, using set notation:

P(A∪B) = P(A) + P(B) – P(AB)

where:

P(A) = probability that event A occurs.

P(B) = probability that event B occurs.

P(A∪B) = probability that event A or event B occurs.

P(A∩B) = probability that event A and event B occur.

For mutually exclusive events, that is events which cannot occur together:

P(A∩B) = 0

The addition rule therefore reduces to:

P(A∪B) = P(A) + P(B)

For independent events, that is events which have no influence on each other:

P(A∩B) = P(A).P(B)

The addition rule therefore reduces to:

P(A∪B) = P(A) + P(B) – P(A).P(B)

Example

Suppose we wish to find the probability of drawing either a king or a spade in a single draw from a pack of 52 playing cards.

We define the events A = 'draw a king' and B = 'draw a spade'

Since there are 4 kings in the pack and 13 spades, but 1 card is both a king and a spade, we have:

P(A∪B) = P(A) + P(B) – P(AB)

$$= \frac{4}{52} + \frac{13}{52} - \frac{1}{52}$$

$$= \frac{16}{52}$$

So, the probability of drawing either a king or a spade is:

$$\frac{16}{52}$$

Additive models Additive models represent a generalization of Multiple Regression (which is a special case of general linear models). Specifically, in linear regression, a linear least-squares fit is computed for a set of predictor or X variables, to predict a dependent Y variable. The well know linear regression equation with m predictors, to predict a dependent variable Y, can be stated as:

$$Y = b_0 + b_1{*}X_1 + .. b_m{*}X_m$$

Where Y stands for the (predicted values of the) dependent variable, X_1 through X_m represent the m values for the predictor variables, and b_0, and b_1 through b_m are the regression coefficients estimated by multiple regression. A generalization of the multiple regression model would be to maintain the additive nature of the model, but to replace the simple terms of the linear equation $b_i{*}X_i$ with $f_i(X_i)$ where f_i is nonparametric function of the predictor X_i. In other words, instead of a single coefficient for each variable (additive term) in the model, in additive models an unspecified (non-parametric) function is estimated for each predictor, to achieve the best prediction of the dependent variable values.

Additive season, damped trend In this Time Series model, the simple exponential smoothing forecasts are "enhanced" both by a damped trend component (independently smoothed with the single parameter ϕ, this model is an extension of Brown's one-parameter linear model) and an additive seasonal component (smoothed with parameter δ). For example, suppose we wanted to forecast from month to month the number of households that purchase a particular consumer electronics device (e.g., VCR). Every year, the number of households that purchase a VCR will increase, however, this trend will be damped (i.e., the upward trend will slowly disappear) over time as the market becomes saturated. In addition, there will be a seasonal component, reflecting the seasonal changes in consumer demand for VCR's from month to month (demand will likely be smaller in the summer and greater during the December holidays). This seasonal component may be additive, for example, a relatively stable number of additional households may purchase VCR's during the December holiday season. To compute the smoothed values for the first season, initial values for the seasonal components are necessary. Also, to compute the smoothed value (forecast) for the first observation in the series, both estimates of S_0 and T_0 (initial trend) are necessary. By default, these values are computed as:

$$T_0 = (1/\phi)*(M_k-M_1)/[(k-1)*p]$$

Where ϕ is the smoothing parameter

k is the number of complete seasonal cycles

M_k is the mean for the last seasonal cycle

M_1 is the mean for the first seasonal cycle

p is the length of the seasonal cycle

and $S_0 = M_1 - p{*}T_0/2$.

Additive season, exponential trend In this Time Series model, the simple exponential smoothing forecasts are "enhanced" both by an exponential trend component (independently smoothed with parameter γ) and an additive seasonal component (smoothed with parameter δ). For example, suppose we wanted to forecast the monthly revenue for a resort area. Every year, revenue may increase by a certain percentage or factor, resulting in an exponential trend in overall revenue. In addition, there could be an additive seasonal component, for example a particular fixed (and slowly changing) amount of added revenue during the December holidays.

To compute the smoothed values for the first season, initial values for the seasonal components are necessary. Also, to compute the smoothed value (forecast) for the first observation in the series, both estimates of S_0 and T_0 (initial trend) are necessary. By default, these values are computed as:

T_0 = exp((log(M_2) – log(M_1))/p)

Where M_2 is the mean for the second seasonal cycle

M_1 is the mean for the first seasonal cycle

p is the length of the seasonal cycle

and S_0 = exp(log(M_1) – p*log(T_0)/2)

Additive season, linear trend In this Time Series model, the simple exponential smoothing forecasts are "enhanced" both by a linear trend component (independently smoothed with parameter γ) and an additive seasonal component (smoothed with parameter δ). For example, suppose we were to predict the monthly budget for snow-removal in a community. There may be a trend component (as the community grows, there is a steady upward trend for the cost of snow removal from year to year). At the same time, there is obviously a seasonal component, reflecting the differential likelihood of snow during different months of the year. This seasonal component could be additive, meaning that a particular fixed additional amount of money is necessary during the winter months, or multiplicative, that is, given the respective budget figure, it may increase by a factor of, for example, 1.4 during particular winter months.

To compute the smoothed values for the first season, initial values for the seasonal components are necessary. Also, to compute the smoothed value (forecast) for the first observation in the series, both estimates of S_0 and T_0 (initial trend) are necessary. By default, these values are computed as:

T_0 = (M_k-M_1)/((k-1)*p

Where k is the number of complete seasonal cycles

M_k is the mean for the last seasonal cycle

M_1 is the mean for the first seasonal cycle

p is the length of the seasonal cycle

and S_0 = M_1 – T_0/2

Additive season, no trend This Time Series model is partially equivalent to the simple exponential smoothing model; however, in addition, each forecast is "enhanced" by an additive seasonal component that is smoothed independently. This model would, for example, be adequate when computing forecasts for monthly expected amount of rain. The amount of rain will be stable from year to year, or change only slowly. At the same time, there will be seasonal changes ("rainy seasons"), which again may change slowly from year to year.

To compute the smoothed values for the first season, initial values for the seasonal components are necessary. The initial smoothed value S_0 will by default be computed as the mean for all values included in complete seasonal cycles.

Adjusted means These are the means that one would get after removing all differences that can be accounted for by the covariate in an analysis of variance design (see ANOVA).

The general formula is

Y-$bar_{j(adj)}$ = Y-bar_j – b(X-bar_j – X-bar)

where Y-$bar_{j(adj)}$ the adjusted mean of group j;

Y-bar_j the mean of group j before adjustment;

b the common regression coefficient;

X-bar_j the mean of the covariate for group j;

X-bar the grand mean of the covariate.

Affine transformation See transformation.

AID (automatic interaction detection) Is a classification program developed by Morgan & Sonquist (1963) that led to the development of the THAID (Morgan & Messenger, 1973) and CHAID (Kass, 1980) classification tree programs. These programs perform multi-level splits when computing classification trees.

Akaike information criterion (AIC) When a model involving q parameters is fitted to data, the criterion is defined as $-L_q + 2q$ where L_q is the maximized log likelihood. Akaike suggested maximizing the criterion to choose between models with different numbers of parameters. It was originally proposed for time-series models, but is also used in regression. Akaike Information Criterion (AIC) can be used in Generalized Linear/Nonlinear Models (GLZ) when comparing the subsets of effects during best subset regression. Since the evaluation of the score statistic does not require iterative computations, best subset selection based on the score statistic is computationally faster, while the selection based on the AIC statistic usually provides more accurate results.

Aleatory An interpretation of probability concerned with how likely events are to happen.

Algorithm As opposed to heuristics (which contain general recommendations based on statistical evidence or theoretical reasoning), algorithms are completely defined, finite sets of steps, operations, or procedures that will produce a particular outcome. For example, with a few exceptions, all computer programs, mathematical formulas, and (ideally) medical and food recipes are algorithms.

Alpha Also known as size or type-1 error. This is the probability that, according to some null hypothesis, a statistical test will generate a false-positive error : affirming a non-null pattern by chance. Conventional methodology for statistical testing is, in advance of undertaking the test, to set a nominal alpha criterion level (often 0.05). The outcome is classified as showing statistical significance if the actual alpha (probability of the outcome under the null hypothesis) is no greater than this nominal alpha criterion level. This reasoning is applicable for all types of statistical testing, including re-randomisation statistics.

Alphanumeric or string variables Alphanumeric variables are variables whose values may be stored as letters, digits or other characters or a combination of them. Alphanumeric variables are also known as string variables and are not available for arithmetic operations in spss.

Alternative hypothesis The alternative hypothesis, H1, is a statement of what a statistical hypothesis test is set up to establish. For example, in a clinical trial of a new drug, the alternative hypothesis might be that the new drug has a different effect, on average, compared to that of the current drug. We would write H1: the two drugs have different effects, on average. The alternative hypothesis might also be that the new drug is better, on average, than the current drug. In this case we would write H1: the new drug is better than the current drug, on average.

The final conclusion once the test has been carried out is always given in terms of the null hypothesis. We either 'reject H0 in favour of H1' or 'do not reject H0'; we never conclude 'reject H1', or even 'accept H1'.

If we conclude 'do not reject H0', this does not necessarily mean that the null hypothesis is true, it only suggests that there is not sufficient evidence against H0 in favour of H1; rejecting the null hypothesis then, suggests that the alternative hypothesis may be true.

Analysis of variance (ANOVA) A data analytic procedure that partitions the variability into components that are attributed to various factors. The variability that cannot be so explained is called the residual or error variability. The sources of variation are presented as the rows of an analysis of variance table in which the columns represent sums-of-squares, degrees of freedom, and mean squares.

Analyze The goal of the Analyze phase is to identify the root cause(s) of quality problems, and to confirm those causes using the appropriate data analysis tools.

And, conjunction, logical conjunction And is an operation on two logical propositions. If p and q are two propositions, (p **and** q) is a proposition that is true if both p and q are true; otherwise, (p **and** q) is false.

Anderson Darling test The Anderson-Darling procedure is a general test to compare the fit of an observed cumulative distribution function to an expected cumulative distribution function. This test is applicable to complete data sets (without censored observations). The critical values for the Anderson-Darling statistic have been tabulated for sample sizes between 10 and 40; this test is not computed for n less than 10 and greater than 40.

The Anderson-Darling test is used in Weibull and Reliability/Failure Time Analysis; see also, Mann-Scheuer-Fertig Test and Hollander-Proschan Test.

ANSI [American National Standards Institute] This body publishes specifications for a number of standard programming languages. The specifications are generally arranged to concur with those of ISO.

Ante The up-front cost of a bet: the money you must pay to play the game. From Latin for "before."

Antecedent An antecedent variable is one that has a causal influence on another variable.

Append a network A function to allow two neural networks (with compatible output and input layers) to be joined into a single network.

Append cases and/or variables Functions that add new cases (i.e., rows of data) and/or variables (i.e., columns of data) to the end of the data set (the "bottom" or the right hand side, respectively). Cases and Variables can also be inserted in arbitrary locations of the data set.

Applet An applet is a small program that is automatically downloaded from a website to your computer when you visit a particular web page; it allows a page to be interactive—to respond to your input. The applet runs on your computer, not the computer that hosted the web page. These materials contain many applets to illustrate statistical concepts and to help you to analyze data. Many of them are accessible directly from the tools page.

Application programming interface (API) Application Programming Interface is a set of functions that conform to conventions of a particular operating system (e.g., Windows) which allows the user to programmatically access the functionality of another program. For example, the kernel of Statistica Neural Networks can by accessed by other programs packages (e.g., Visual Basic, Statistica basic, Delphi, C, C++) in a variety of ways.

Arrow An element in a path diagram used to indicate causal flow from one variable to another, or, in narrower interpretation, to show which of two variables in a linear equation is the independent variable and which is the dependent variable.

Assignable causes and actions In the context of monitoring quality characteristics you have to distinguish between two different types of variability: Common cause variation describes random variability that is inherent in the process and affects all individual values. Ideally, when your process is in-control, only common cause variation will be present. In a quality control chart, it will show up as a random fluctuation of the individual samples around the center line with all samples falling between the upper and lower control limit and no non-random patterns (runs) of adjacent samples. Special cause or assignable cause variation is due to specific circumstances that can be accounted for. It will usually show up in the QC chart as outlier samples (i.e., exceeding the lower or upper control limit) or as a systematic pattern (run) of adjacent samples. It will also affect the calculation of the chart specifications (center line and control limits).

With some software programs, if you investigate the out-of-control conditions and you find an explanation for them, you can assign descriptive labels to those out-of-control samples and explain the causes (e.g., valve defect) and actions that have been taken (e.g. valve fixed). Having causes and actions displayed in the chart will document that the center line and the control limits of the chart are affected by special cause variation in the process.

Association Two variables are associated if some of the variability of one can be accounted for by the other. In a scatterplot of the two variables, if the scatter in the values of the variable plotted on the vertical axis is smaller in narrow ranges of the variable plotted on the horizontal axis (*i.e.,* in vertical "slices") than it is overall, the two variables are associated. The correlation coefficient is a measure of linear association, which is a special case of association in which large values of one variable tend to occur with large values of the other, and small values of one tend to occur with small values of the other (positive association), or in which large values of one tend to occur with small values of the other, and *vice versa* (negative association).

Association rules Data mining for association rules is often the first and most useful method for analyzing data that describe transactions, lists of items, unique phrases (in text mining) etc. In general, association rules take the form If Body then Head, where Body and Head stand for simple codes, text values, items, consumer choices, phrases etc., or the conjunction of codes and text values etc. (e.g., if (Car=Porsche and Age<20 and ThrillSeeking=High) then (Risk=High and Insurance=High); here the logical conjunction before the then would be the Body, and the logical conjunction following the then would be the Head of the association rule). The A-priori algorithm is a popular and efficient algorithm for deriving such association rules from large data sets, based on some user-defined "threshold" values for rule.

Asymmetric A characteristic of a measure of association. A measure is asymmetric when the value it takes depends on which variable is considered to be the independent and which the dependent variable.

Asymmetrical distribution If you split the distribution in half at its mean (or median), then the distribution of values on the two sides of this central point would not be the same (i.e., not symmetrical) and the distribution would be considered "skewed."

Attribute (attribute variable) An alternative name for a nominal variable.

Augmented product moment matrix For a set of p variables, this is a (p + 1) X (p + 1) square matrix. The first p rows and columns contain the matrix of moments about zero, while the last row and column contain the sample means for the p variables. The matrix is therefore of the form:

$$M_A = \begin{bmatrix} M & \bar{x} \\ \bar{x}' & 1 \end{bmatrix}$$

where M is a matrix with element

$$M_{jk} = \frac{1}{N}\sum_{i-1}^{N} X_{ij} X_{ik}$$

and $\bar{x}$ is a vector with the means of the variables (see Structural Equation Modeling).

Autoassociative network A neural network (usually a multilayer perceptron) designed to reproduce its inputs at its outputs, while "squeezing" the data through a lower-dimensionality middle layer. Used for compression or dimensionality reduction purposes.

Autocorrelation Autocorrelation is the correlation (relationship) between members of a time series of observations, such as weekly share prices or interest rates, and the same values at a fixed time interval later.

More technically, autocorrelation occurs when residual error terms from observations of the same variable at different times are correlated (related).

Automatic network designer An heuristic algorithm (implemented in Statistica Neural Networks) which experimentally determines an appropriate network architecture to fit a specified data set.

Average *Average* usually denotes the arithmetic mean, but it can also denote the median, the mode, the geometric mean, and weighted means, among other things.

Axioms of probability There are three axioms of probability: 1. Chances are always at least zero.

2. The chance that *something* happens is 100%. (3) If two events cannot both occur at the same time (if they are disjoint or mutually exclusive), the chance that either one occurs is the sum of the chances that each occurs. For example, consider an experiment that consists of tossing a coin once. The first axiom says that the chance that the coin lands heads, for instance, must be at least zero. The second axiom says that the chance that the coin either lands heads or lands tails or lands on its edge or doesn't land at all is 100%. The third axiom says that the chance that the coin either lands heads or lands tails is the sum of the chance that the coin lands heads and the chance that the coin lands tails, because both cannot occur in the same coin toss. All other mathematical facts about probability can be derived from these three axioms. For example; it is true that the chance that an event does not occur is (100% – the chance that the event occurs). This is a consequence of the second and third axioms.

B

B coefficients A line in a two dimensional or two-variable space is defined by the equation Y=a+b*X; in full text: the Y variable can be expressed in terms of a constant (a) and a slope (b) times the X variable. The constant is also referred to as the intercept, and the slope as the regression coefficient or B coefficient. In general then, multiple regression procedures will estimate a linear equation of the form:

$Y = a + b_1{*}X_1 + b_2{*}X_2 + ... +b_p{*}X_p$

Note that in this equation, the regression coefficients (or B coefficients) represent the independent contributions of each independent variable to the prediction of the dependent variable. However, their values may not be comparable between variables because they depend on the units of measurement or ranges of the respective variables. Some software products will produce both the raw regression coefficients (B coefficients) and the Beta coefficients (note that the Beta coefficients are comparable across variables).

Back propagation Back propagation is the best known training algorithm for neural networks and still one of the most useful. Devised independently by Rumelhart et. al. (1986), Werbos (1974), and Parker (1985), it is thoroughly described in most neural network textbooks (e.g., Patterson, 1996; Fausett, 1994; Haykin, 1994). It has lower memory requirements than most algorithms, and usually reaches an acceptable error level quite quickly, although it can then be very slow to converge properly on an error minimum. It can be used on most types of networks, although it is most appropriate for training multilayer perceptrons.

Back propagation includes:

Time-dependent learning rate.

Time-dependent momentum rate.

Random shuffling of order of presentation.

Additive noise during training.

Independent testing on selection set.

A variety of stopping conditions.

RMS error plotting.

Selectable error function.

The last five bulleted items are equally available in other iterative algorithms, including conjugate gradient descent, Quasi-Newton, Levenberg-Marquardt, quick propagation, Delta-bar-Delta, and Kohonen training (apart from noise in conjugate gradients, Kohonen and Levenberg-Marquardt, and selectable error function in Levenberg-Marquardt).

Backward elimination 2 A procedure for exploratory analysis that successively deletes variables from the regression equation until there is a significant reduction in the variance explained.

Bagging (voting, averaging) The concept of bagging (voting for classification, averaging for regression-type problems with continuous dependent variables of interest) applies to the area of predictive data mining, to combine the predicted classifications (prediction) from multiple models, or from the same type of model for different learning data. It is also used to address the inherent instability of results when applying complex models to relatively small data sets.

Suppose your data mining task is to build a model for predictive classification, and the dataset from which to train the model (learning data set, which contains observed classifications) is relatively small. You could repeatedly sub-sample (with replacement) from the dataset, and apply, for example, a tree classifier (e.g., C&RT and CHAID) to the successive samples. In practice, very different trees will often be grown for the different samples, illustrating the instability of models often evident with small datasets. One method of deriving a single prediction (for new observations) is to use all trees found in the different samples, and to apply some simple voting: The final classification is the one most often predicted by the different trees. Note that some weighted combination of predictions (weighted vote, weighted average) is also possible, and commonly used. A sophisticated (machine learning) algorithm for generating weights for weighted prediction or voting is the Boosting procedure.

Balanced ANOVA design An ANOVA design is a balanced ANOVA design when all cells in the ANOVA design have equal N, when there are no missing cells in the design, and, if nesting is present, when the nesting is balanced so that equal numbers of levels of the factors that are nested appear in the levels of the factor(s) that they are nested in. Most between-groups ANOVA designs can be analyzed much more efficiently when they are balanced ANOVA designs.

Balanced cross over design See cross-over design

Bar chart A bar chart is a way of summarising a set of categorical data. It is often used in exploratory data analysis to illustrate the major features of the distribution of the data in a convenient form. It displays the data using a number of rectangles, of the same width, each of which represents a particular category. The length (and hence area) of each rectangle is proportional to the number of cases in the category it represents, for example, age group, religious affiliation.

Bar charts are used to summarise nominal or ordinal data.

Bar charts can be displayed horizontally or vertically and they are usually drawn with a gap between the bars (rectangles), whereas the bars of a histogram are drawn immediately next to each other.

Bar dev plot The "bar deviation" plot is similar to the Bar X plot, in that individual data points are represented by vertical bars, however, the bars connect the data points to a user-selectable baseline. If the baseline value is different than the plot's Y-axis minimum, then individual bars will extend either up or down, depending on the direction of the "deviation" of individual data points from the baseline.

Bar left y plot In this plot, one horizontal bar is drawn for each data point (i.e., each pair of XY coordinates), connecting the data point and the left Y-axis. The vertical position of the bar is determined by the data point's Y value, and its length by the respective X value.

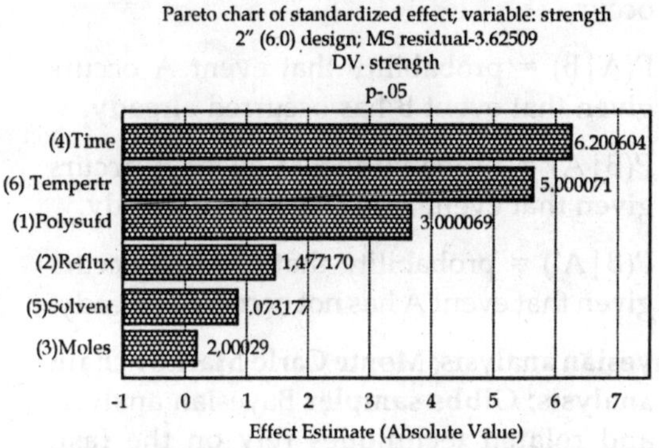

Bar right y plot In this plot, one horizontal bar is drawn for each data point (i.e., each pair of XY coordinates), connecting the data point and the right Y-axis. The vertical position of the bar is determined by the data point's Y value, and its length by the respective X value.

Bar top plot (Also known as "hanging" column plots.) In this plot, one vertical bar is drawn for each data point (i.e., each pair of XY coordinates), connecting the data point and the upper X-axis. The horizontal position of the bar is determined by the data point's X value, and its length by the respective Y value.

Bar x plot In this plot, one vertical bar is drawn for each data point (i.e., each pair of XY coordinates), connecting the data point and the lower X-axis.

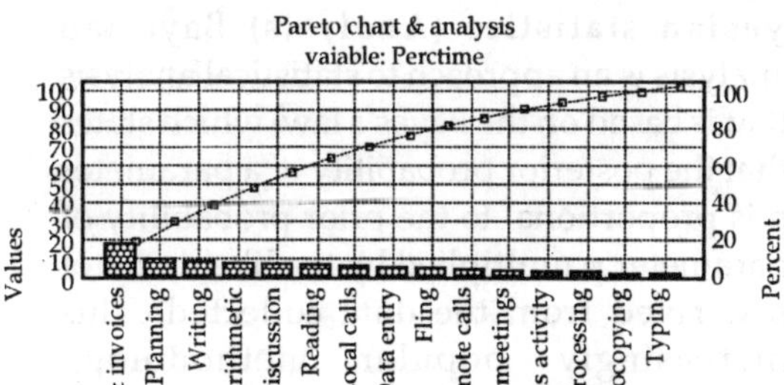

The horizontal position of the bar is determined by the data point's X value, and its height by the respective Y value.

Bar/column plots, 2D The Bar/Column Plot represents sequences of values as bars or columns (one case is represented by one bar/column). If more than one variable is selected, each plot can be represented in a separate graph or all of them can be combined in one display as multivariate clusters of bars/columns.

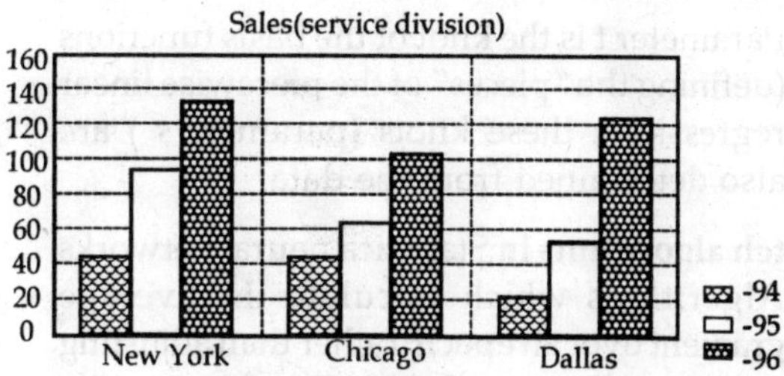

Bartlett window In Time Series, the Bartlett window is a weighted moving average transformation used to smooth the periodogram values. In the Bartlett window (Bartlett, 1950) the weights are computed as:

$w_j = 1-(j/p)$ (for j = 0 to p)

$w_{-j} = w_j$ (for $j \neq 0$)

where $p = (m-1)/2$

This weight function will assign the greatest weight to the observation being smoothed in the center of the window, and increasingly smaller weights to values that are further away from the center.

Basis functions Basis functions of predictor variables (X) play an important role in the estimation of Multivariate Adaptive Regression Splines (MARSplines). Specifically, marsplines uses two-sided truncated functions of the form $(x-t)_+$ as basis functions for linear or non-linear expansion which approximates the relationships between the response and predictor variables.

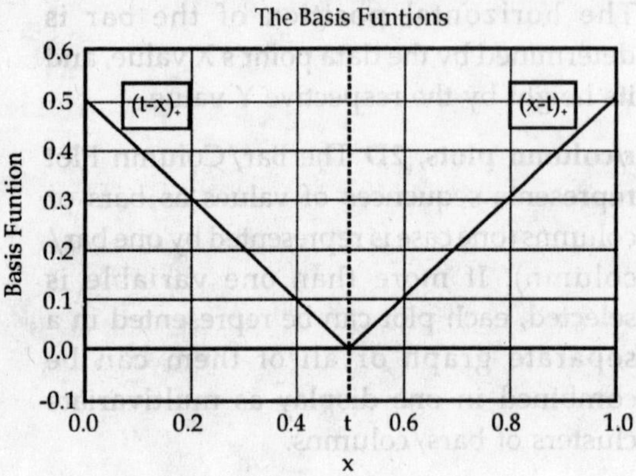

Parameter t is the knot of the basis functions (defining the "pieces" of the piecewise linear regression); these knots (parameters) are also determined from the data.

Batch algorithms in Statistica neural networks Algorithms which calculate the average gradient over an epoch, rather than adjusting on a case-by-case basis during training. Quick propagation, Delta-Bar-Delta, conjugate gradient descent and Levenberg-Marquardt are all batch algorithms.

Bayes' rule Bayes' rule expresses the conditional probability of the event A given the event B in terms of the conditional probability of the event B given the event A:

$$P(A \mid B) = P(B \mid A) \times P(A)/(P(B \mid A) \times P(A) + P(B \mid A^c) \times P(A^c))$$

Bayes' theorem Bayes' Theorem is a result that allows new information to be used to update the conditional probability of an event.

Using the multiplication rule:

$$P(A|B) = \frac{P(A \cap B)}{P(B)} = \frac{P(B|A).P(A)}{P(B)}$$

This is Bayes' Theorem in its simplest form. Using the Law of Total Probability:

$$P(A|B) = \frac{P(B|A).P(A)}{P(B|A).P(A) + P(B|A').P(A')}$$

where:

P(A) = probability that event A occurs.

P(B) = probability that event B occurs.

P(A′) = probability that event A does not occur.

P(A | B) = probability that event A occurs given that event B has occurred already.

P(B | A) = probability that event B occurs given that event A has occurred already.

P(B | A′) = probability that event B occurs given that event A has not occurred already.

Bayesian analysis; Monte Carlo Markov chain analysis; Gibbs sampler Bayesian analysis and related techniques rely on the fast, efficient, and "precise" generation of random numbers from a variety of distributions. Numerous different types of specialized libraries are available for this purpose, and a good deal of ongoing research in this area is further refining these techniques and their applications.

Bayesian inference Also called subjectivist.

An inferential paradigm that interprets observed data by using Bayes' theorem to transform prior (epistemic) probabilities into posterior probabilities.

Bayesian networks Networks based on Bayes' theorem, on the inference of probability distributions from data sets.

Bayesian statistics (Analysis) Bayesian analysis is an approach to statistical analysis that is based on the Bayes's law, which states that the posterior probability of a parameter p is proportional to the prior probability of parameter p multiplied by the likelihood of p derived from the data collected. This increasingly popular methodology represents an alternative to the traditional (or frequentist probability) approach: whereas the latter attempts to establish confidence intervals around parameters, and/or falsify a-priori null-hypotheses, the Bayesian approach attempts to keep track of how a-priori expectations about some phenomenon of interest can be refined, and how observed data can be integrated with such a-priori beliefs, to arrive at updated posterior expectations about the phenomenon.

A good metaphor (and actual application) for the Bayesian approach is that of a physician who applies consecutive examinations to a patient so as to refine the certainty of a particular diagnosis: The results of each individual examination or test should be combined with the a-priori knowledge about the patient, and expectation that the respective diagnosis is correct. The goal is to arrive at a final diagnosis which the physician believes to be correct with a known degree of certainty.

Behaviourist inference See frequentist inference.

Bell shaped curve Symmetrical, single-peaked frequency distribution.

Also called the normal curve or gaussian curve.

Bernouilli process This is the simplest probability model - a single trial between two possible outcomes such as a coin toss. The distribution depends upon a single parameter,'p', representing the probability attributed to one defined outcome out of the two possible outcomes. Also see : binomial distribution, poisson process.

Bernoulli distribution The Bernoulli distribution best describes all situations where a "trial" is made resulting in either "success" or "failure," such as when tossing a coin, or when modeling the success or failure of a surgical procedure.

Bernoulli trials A sequence of independent events all having the same probability. The events and their complements are usually called success and failure, respectively.

Best network retention A facility (implemented in statistica Neural Networks) to automatically store the best neural network discovered during training, for later restoration at the end of a set of experiments.

Best subset regression A model-building technique which finds subsets of predictor variables that best predict responses on a dependent variable by linear (or nonlinear) regression.

BETA Also known as type-2 error, BETA is the complement to power : BETA = (1-POWER). This is the probability that a statistical test will generate a false-negative error : failing to assert a defined pattern of deviation from a null pattern in circumstances where the defined pattern exists. Conventional methodology for statistical testing is to set in advance a nominal alpha criterion level - the corresponding level for BETA will depend upon the nominal alpha criterion level and upon further considerations including the strength of the pattern in the data and the sample size. Interest is generally in the relative power of different tests rather than in an absolute value.

Beta coefficients The Beta coefficients are the regression coefficients you would have obtained had you first standardized all of your variables to a mean of 0 and a standard deviation of 1. Thus, the advantage of Beta coefficients (as compared to B coefficients which are not standardized) is that the magnitude of these Beta coefficients allow you to compare the relative contribution of each independent variable in the prediction of the dependent variable.

Beta distribution The beta distribution (the term first used by Gini, 1911) is defined as:

$f(x) = \Gamma(\nu+\omega)/(\Gamma()\Gamma(\omega)) * x^{\nu-1} * (1-x)^{\omega-1}$

$0 \le x \le 1$

$\nu > 0, \omega > 0$

where Γ (gamma) is the Gamma function
ν, ω are the shape parameters.

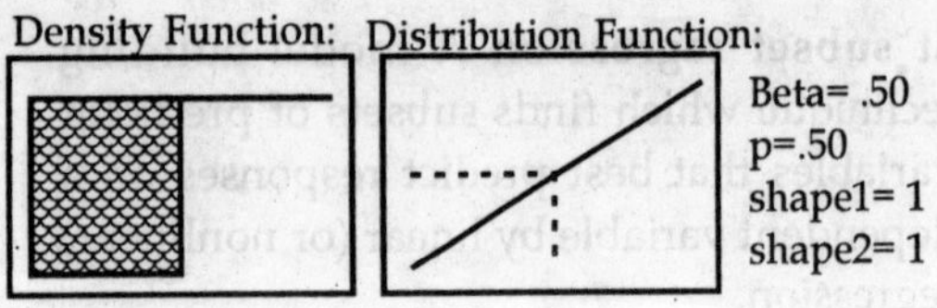

Between subject effects A comparison of experimental results in which the attributes being compared belong to different subjects. cf. within-subject effects.

Bias Bias is a term which refers to how far the average statistic lies from the parameter it is estimating, that is, the error which arises when estimating a quantity. Errors from chance will cancel each other out in the long run, those from bias will not.

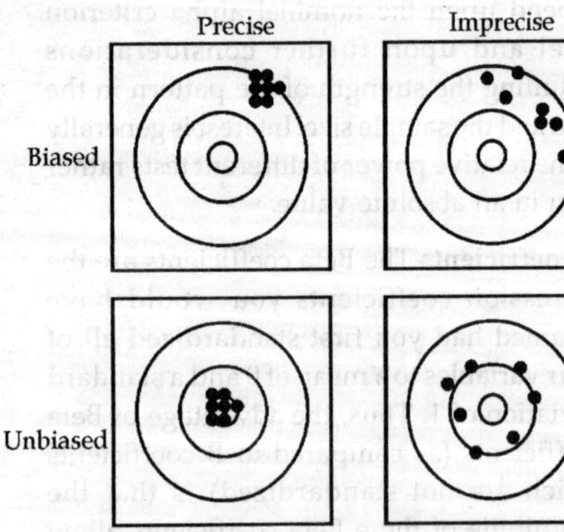

The following illustrates bias and precision, where the target value is the bullseye:

Example The police decide to estimate the average speed of drivers using the fast lane of the motorway and consider how it can be done. One method suggested is to tail cars using police patrol cars and record their speeds as being the same as that of the police car. This is likely to produce a biased result as any driver exceeding the speed limit will slow down on seeing a police car behind them. The police then decide to use an unmarked car for their investigation using a speed gun operated by a constable. This is an unbiased method of measuring speed, but is imprecise compared to using a calibrated speedometer to take the measurement.

Bimodal Curve with two equal scores of highest frequency.

Bimodal distribution A distribution that has two modes (thus two "peaks").

Bimodality of the distribution in a sample is often a strong indication that the distribution of the variable in population is not normal. Bimodality of the distribution may provide important information about the nature of the investigated variable (i.e., the measured quality). For example, if the variable represents a reported preference or attitude, then bimodality may indicate a polarization of opinions. Often however, the bimodality may indicate that the sample is not homogenous and the observations come in fact from two or more "overlapping" distributions. Sometimes, bimodality of the distribution may indicate problems with the measurement instrument (e.g, "gage calibration problems" in natural sciences, or "response biases" in social sciences).

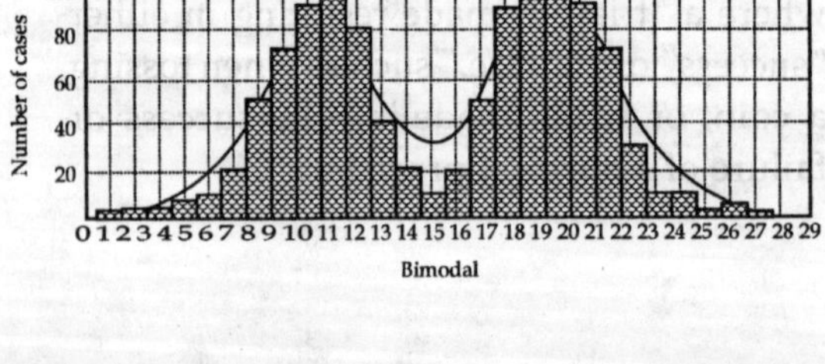

See also unimodal distribution, multimodal distribution.

Bin See class interval.

Binary variable Also called *dichotomous* variable

A categorical variable which has exactly two possible values.

Binomial Event with only two possible outcomes.

Binomial coefficient See combinations.

Binomial distribution Binomial distributions model (some) discrete random variables.

Typically, a binomial random variable is the number of successes in a series of trials, for example, the number of 'heads' occurring when a coin is tossed 50 times.

A discrete random variable X is said to follow a Binomial distribution with parameters n and p, written $X \sim Bi(n,p)$ or $X \sim B(n,p)$, if it has probability distribution:

$$P(X=x) = \binom{n}{x} p^x (1-p)^{n-x}$$

where:

$x = 0, 1, 2, \ldots\ldots, n$

$n = 1, 2, 3, \ldots\ldots$

p = success probability; $0 < p < 1$, and:

$$\binom{n}{x} = \frac{n!}{x!\,(n-x)!}$$

The trials must meet the following requirements:

a) the total number of trials is fixed in advance;

b) there are just two outcomes of each trial; success and failure;

c) the outcomes of all the trials are statistically independent;

d) all the trials have the same probability of success.

The Binomial distribution has expected value $E(X) = np$ and variance $V(X) = np(1-p)$.

Binomial probability distribution For binomial events, the frequency of the number of favorable outcomes. For a large number of trials, the binomial distribution approaches the normal distribution.

Binomial test This is a statistical test referring to a repeated binary process such as would be expected to generate outcomes with a binomial distribution. A value for the parameter 'p' is hypothesised (null hypothesis) and the difference of the actual value from this is assessed as a value of alpha.

Bivariate analysis Statistical analysis involving two variables

Bivariate Having or having to do with two variables. For example, bivariate data are data where we have two measurements of each "individual." These measurements might be the heights and weights of a group of people (an "individual" is a person), the heights of fathers and sons (an "individual" is a father-son pair), the pressure and temperature of a fixed volume of gas (an "individual" is the volume of gas under a certain set of experimental conditions), *etc.* Scatterplots, the correlation coefficient, and regression make sense for bivariate data but not univariate data. *C.f.* univariate.

Bivariate analysis Statistical analysis involving two variables

Bivariate normal distribution Two variables follow the bivariate normal distribution if for each value of one variable, the corresponding values of another variable are normally distributed. The bivariate normal probability distribution function for a pair of continuous random variables (X and Y) is given by:

$f(x,y) = \{1/[2\pi\sigma_1\sigma_2 * (1-\rho)^{1/2}]\} * \exp\{[-1/2(1-\rho^2)] * \{[(x-\mu_1)/\rho^1]^2 -$

$\rho\,[(x-\mu_1)/\sigma_1] * [(y-\mu_2)/\sigma_2] + [(y-\mu_2)/\sigma_2]^2]\}$
$-\infty < x < \infty, -\infty < y < \infty, -\infty < \mu_1 < \infty, -\infty < \mu_2 < \infty, \sigma_1 > 0, \sigma_2 > 0$, and $-1 < \rho < 1$

Where μ_1, μ_2 are the respective means of the random variables X and Y

σ_1, σ_2 are the respective standard deviations of the random variables X and Y

ρ is the correlation coefficient of X and Y

e is the base of the natural logarithm, sometimes called Euler's e

π is the constant Pi (3.14...)

See also, Normal Distribution, Elementary Concepts (Normal Distribution).

Bivariate relationship A statistical relationship between two variables.

Blind, blind experiment In a blind experiment, the subjects do not know whether they are in the treatment group or the control group. In order to have a blind experiment with human subjects, it is usually necessary to administer a placebo to the control group.

Blocking (in experimental design) In some experiments, observations are organized in natural "chunks" or blocks. You want to make sure that these blocks do not bias your estimates of main effects or interactions. For example, consider an experiment to improve the quality of special ceramics, produced in a kiln. The size of the kiln is limited so that you cannot produce all runs (observations) of your experiment at once. In that case you need to break up the experiment into blocks. However, you do not want to run positive factor settings (for all factors in your experiment) in one block, and all negative settings in the other. Otherwise, any incidental differences between blocks would systematically affect all estimates of the main effects and interactions of the factors of interest. Rather, you want to distribute the runs over the blocks so that any differences between blocks (i.e., the blocking factor) do not bias your results for the factor effects of interest. This is accomplished by treating the blocking factor as another factor in the design. Blocked designs often also have the advantage of being statistically more powerful, because they allow you to estimate and control the variability in the production process that is due to differences between blocks.

Bonferroni adjustment When performing multiple statistical significance tests on the same data, the Bonferroni adjustment can be applied to make it more "difficult" for any one test to be statistically significant. For example, when reviewing multiple correlation coefficients form a correlation matrix, accepting and interpreting the correlations that are statistically significant at the conventional .05 level may be inappropriate, given that multiple tests are performed. Specifically, the alpha error probability of erroneously accepting the observed correlation coefficient as not-equal-to-zero when in fact (in the population) it is equal to zero may be much larger than .05 in this case.

The Bonferroni adjustment usually is accomplished by dividing the alpha level (usually set to .05, .01, etc.) by the number of tests being performing. For instance, suppose you performed multiple tests of individual correlations from the same correlation matrix. The Bonferroni adjusted level of significance for any one correlation would be:

$.05 / 5 = .01$

Any test that results in a p-value of less than .01 would be considered statistically significant; correlations with a probability value greater than .01 (including those with p-values between .01 and .05) would be considered non-significant.

Bonferroni test This post hoc test can be used to determine the significant differences between group means in an analysis of variance setting. The Bonferroni test is very conservative when a large number of group means are being compared .

Boolean Referring to the operations of union, intersection, and complementation, their relationships and representations.

Boosting The concept of boosting applies to the area of predictive data mining, to generate multiple models or classifiers (for prediction or classification), and to derive weights to combine the predictions from those models into a single prediction or predicted classification.

A simple algorithm for boosting works like this: Start by applying some method (e.g., a tree classifier such as C&RT or CHAID) to the learning data, where each observation is assigned an equal weight. Compute the predicted classifications, and apply weights to the observations in the learning sample that are inversely proportional to the accuracy of the classification. In other words, assign greater weight to those observations that were difficult to classify (where the misclassification rate was high), and lower weights to those that were easy to classify (where the misclassification rate was low). In the context of C&RT for example, different misclassification costs (for the different classes) can be applied, inversely proportional to the accuracy of prediction in each class. Then apply the classifier again to the weighted data (or with different misclassification costs), and continue with the next iteration (application of the analysis method for classification to the re-weighted data).

Boosting will generate a sequence of classifiers, where each consecutive classifier in the sequence is an "expert" in classifying observations that were not well classified by those preceding it. During deployment (for prediction or classification of new cases), the predictions from the different classifiers can then be combined (e.g., via voting, or some weighted voting procedure) to derive a single best prediction or classification.

Note that boosting can also be applied to learning methods that do not explicitly support weights or misclassification costs. In that case, random sub-sampling can be applied to the learning data in the successive steps of the iterative boosting procedure, where the probability for selection of an observation into the subsample is inversely proportional to the accuracy of the prediction for that observation in the previous iteration (in the sequence of iterations of the boosting procedure).

Bootstrap This is a form of randomisation test which is one of the alternatives to exhaustive re-randomisation. The bootstrap scheme involves generating subsets of the data on the basis of random sampling with replacements as the data are sampled. Such resampling provides that each datum is equally represented in the randomisation scheme; however, the bootstrap procedure has features which distinguish it from the procedure of a monte-carlo test. The distinguishing features of the bootstrap procedure are concerned with over-sampling - there is no constraint upon the number of times that a datum may be represented in generating a single resampling subset; the size of the resampling subsets may be fixed arbitrarily independently of the parameter values of the experimental design and may even exceed the total number of data. The positive motive for bootstrap resampling is the general relative ease of devising an appropriate resampling algorithm(1) when the experimental design is novel or complex. A negative aspect of the bootstrap is that the form of the resampling distribution with prolonged resampling converges to a form which depends not only upon the data and the test statistic, but also upon the bootstrap resampling subset size - thus the resampling distribution should not be expected to converge to the gold standard(1) form of the exact test as is the case for monte-carlo resampling. An effective necessity for the bootstrap procedure is a source of random codes or an effective pseudo-random generator.

Bootstrap estimate of standard error The name for this idea comes from the idiom "to pull oneself up by one's bootstraps," which connotes getting out of a hole without anything to stand on. The idea of the bootstrap is to assume, for the purposes of estimating uncertainties, that the sample is the population, then use the SE for sampling from the sample to estimate the SE of sampling from the population. For sampling from a box of numbers, the SD of the sample is the bootstrap estimate of the SD of the box from which the sample is drawn. For sample percentages, this takes a particularly simple form: the SE of the sample percentage of n draws from a box, with replacement, is SD(box)/$n^{1/2}$, where for a box that contains only zeros and ones, SD(box) = ((fraction of ones in box)×(fraction of zeros in box))$^{1/2}$. The bootstrap estimate of the SE of the sample percentage consists of estimating SD(box) by ((fraction of ones in sample)×(fraction of zeros in sample))$^{1/2}$. When the sample size is large, this approximation is likely to be good.

Boundary case A boundary case occurs when a parameter iterates to the "boundary" of the permissible "parameter space". For example, a variance can only take on values from 0 to infinity. If, during iteration, the program attempts to move an estimate of a variance below zero, the program will constrain it to be on the boundary value of 0.

For some problems (for example a Heywood Case in factor analysis), it may be possible to reduce the discrepancy function by estimating a variance to be a negative number. In that case, the program does "the best it can" within the permissible parameter space, but does not actually obtain the "global minimum" of the discrepancy function.

Box and whisker plot (or boxplot) A box and whisker plot is a way of summarising a set of data measured on an interval scale. It is often used in exploratory data analysis. It is a type of graph which is used to show the shape of the distribution, its central value, and variability. The picture produced consists of the most extreme values in the data set (maximum and minimum values), the lower and upper quartiles, and the median.

A box plot (as it is often called) is especially helpful for indicating whether a distribution is skewed and whether there are any unusual observations (outliers) in the data set.

Box and whisker plots are also very useful when large numbers of observations are involved and when two or more data sets are being compared.

Box ljung q statistic In Time Series analysis, you can shift a series by a given lag k. For that given lag, the Box-Ljung Q statistic is defined by:

$Q_k = n*(n+2)*Sum(r_i^2/(n-1))$

for i = 1 to k

When the number of observations is large, then the Q statistic has a Chi- square distribution with k-p-q degrees of freedom, where p and q are the number of autoregressive and moving average parameters, respectively.

Box model An analogy between an experiment and drawing numbered tickets "at random" from a box with replacement. For example, suppose we are trying to evaluate a cold remedy by giving it or a placebo to a group of n individuals, randomly choosing half the individuals to receive the remedy and half to receive the placebo. Consider the median time to recovery for all the individuals (we assume everyone recovers from the cold eventually; to simplify things, we also assume that no one recovered in exactly the median time, and that n is even). By definition, half the individuals got better in less than the median time, and half in more than the median time. The individuals who received the treatment are a random sample of size $n/2$ from the set of n subjects, half of whom got better in less than median time, and half in longer than median time. If the remedy is ineffective, the number of subjects who received the remedy and who recovered in less than median time is like the sum of $n/2$ draws with replacement from a box with two tickets in it: one with a "1" on it, and one with a "0" on it.

Box plot/means (block stats graphs) This type of Block Stats Graph will produce a box plot of means (and standard errors and standard deviations) for the columns or rows of the block. Each box will represent data from one column or row.

Box plot/medians (block stats graphs) This type of Block Stats Graph will produce a box plot of medians (and min/max values and 25th and 75th percentiles) for the columns or rows of the block. Each box will represent data from one column or row.

Box plots 3d border style ranges In this style of 3D Sequential Box Plot, the ranges of values of selected variables are plotted separately for groups of cases defined by values of a categorical (grouping) variable. The central tendency (e.g., median or mean), and range or variation statistics (e.g., quartiles, standard errors, or standard deviations) are computed for each variable and for each group of cases and the selected values are presented as points with "whiskers," and the ranges marked by the "whiskers" are connected with lines (i.e., range borders) separately for each variable.

3D Range plots differ from 3D Box plots in that for Range plots, the ranges are the values of the selected variables (e.g., one variable contains the minimum range values and another variable contains the maximum range values) while for Box plots, the ranges are calculated from variable values (e.g., standard deviations, standard errors, or min-max value).

Box plots 3d double ribbon ranges In this style of 3D Sequential Box Plot, the ranges of values of selected variables are plotted separately for groups of cases defined by values of a categorical (grouping) variable. The range or variation statistics (e.g., quartiles, standard errors, or standard deviations) are computed for each variable and for each group of cases and the selected values are presented as double ribbons.

3D Range plots differ from 3D Box plots in that for Range plots, the ranges are the values of the selected variables (e.g., one variable contains the minimum range values and another variable contains the maximum range values) while for Box plots the ranges are calculated from variable values (e.g., standard deviations, standard errors, or min-max value).

Box plots 3d flying blocks In this style of 3D Sequential Box Plot, the ranges of values of selected variables are plotted separately for groups of cases defined by values of a categorical (grouping) variable. The central tendency (e.g., median or mean), and range or variation statistics (e.g., quartiles, standard errors, or standard deviations) are computed for each variable and for each group of cases and the selected values are presented as "flying" blocks.

3D Range plots differ from 3D Box plots in that for Range plots, the ranges are the values of the selected variables (e.g., one variable contains the minimum range values and another variable contains the maximum range values) while for Box plots the ranges are calculated from variable values (e.g., standard deviations, standard errors, or min-max value).

Box plots 3d flying boxes In this style of 3D Sequential Box Plot, the ranges of values of selected variables are plotted separately for groups of cases defined by values of a categorical (grouping) variable. The central tendency (e.g., median or mean), and range or variation statistics (e.g., quartiles, standard errors, or standard deviations) are computed for each variable and for each group of cases and the selected values are presented "flying" boxes.

3D Range plots differ from 3D Box plots in that for Range plots, the ranges are the values of the selected variables (e.g., one variable contains the minimum range values and another variable contains the maximum range values) while for Box plots the ranges are calculated from variable values (e.g., standard deviations, standard errors, or min-max value).

Box plots 3d points In this style of 3D Sequential Box Plot, the ranges of values of selected variables are plotted separately for groups of cases defined by values of a categorical (grouping) variable. The central tendency (e.g., median or mean), and range or variation statistics (e.g., quartiles, standard errors, or standard deviations) are computed for each variable and for each group of cases and the selected values are presented as point markers connected by a line.

3D Range plots differ from 3D Box plots in that for Range plots, the ranges are the values of the selected variables (e.g., one variable contains the minimum range values and another variable contains the maximum range values) while for Box plots, the ranges are calculated from variable values (e.g., standard deviations, standard errors, or min-max value).

Box plots, 2d box whiskers This type of box plot will place a box around the midpoint (i.e., mean or median) which represents a selected range (i.e., standard error, standard deviation, min-max, or constant) and whiskers outside of the box which also represent a selected range.

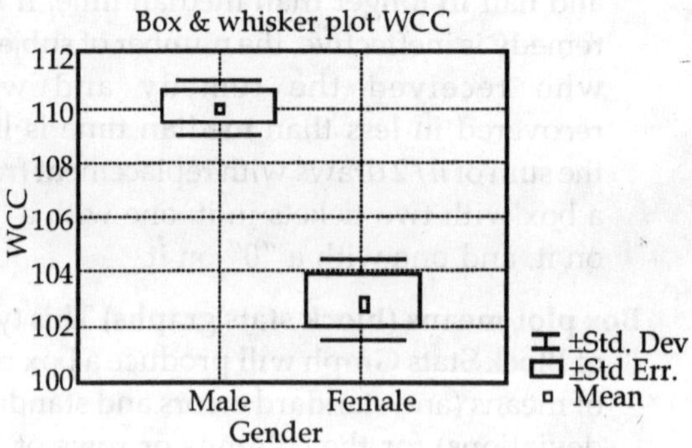

Box plots, 2d boxes This type of box plot will place a box around the midpoint (i.e., mean or median) which represents the selected range (i.e., standard error, standard deviation, min-max, or constant).

Box plots, 2d In Box Plots (this term was first used by Tukey, 1970), ranges or distribution characteristics of values of a selected variable (or variables) are plotted separately for groups of cases defined by values of a categorical (grouping) variable. The central tendency (e.g., median or mean), and range or variation statistics (e.g., quartiles, standard errors, or standard deviations) are computed for each group of cases and the selected values are presented in the selected box plot style. Outlier data points can also be plotted.

Box plots, 2d whiskers In this style of box plot, the range (i.e., standard error, standard deviation, min-max, or constant) is represented by "whiskers" (i.e., as a line with a serif on both ends.

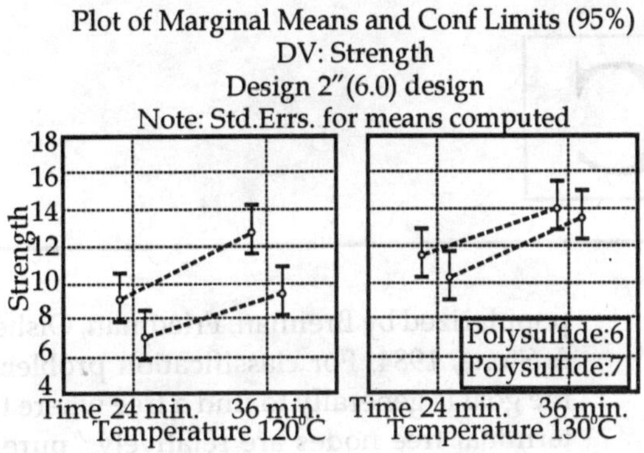

Box plots, 3d In Box Plots (this term was first used by Tukey, 1970), ranges or distribution characteristics of values of selected variables are plotted separately for groups of cases defined by values of a categorical (grouping) variable. The central tendency (e.g., median or mean), and range or variation statistics (e.g., quartiles, standard errors, or standard deviations) are computed for each group of cases and the selected values are presented in the selected box plot style. Outlier data points can also be plotted.

Branch and bound Exploration of a randomisation distribution in such a way as to anticipate the effect of the next randomisation(3) relative to the present randomisation(3). This allows selective search of particular zones of a randomisation distribution; in the context of a randomisation test such selective search may be concerned with the tail of the randomisation distribution.

Breakdown point The breakdown point of an estimator is the smallest fraction of observations one must corrupt to make the estimator take any value one wants.

Breakdowns Breakdowns are procedures which allow us to calculate descriptive statistics and correlations for dependent variables in each of a number of groups defined by one or more grouping (independent) variables. It is used as either a hypothesis testing or exploratory method.

Brushing Perhaps the most common and historically first widely used technique explicitly identified as graphical exploratory data analysis is brushing, an interactive method allowing one to select on-screen specific data points or subsets of data and identify their (e.g., common) characteristics, or to examine their effects on relations between relevant variables (e.g., in scatterplot matrices) or to identify (e.g., label) outliers.

Burt table Multiple correspondence analysis expects as input (i.e., the program will compute prior to the analysis) a so-called Burt table. The Burt table is the result of the inner product of a design or indicator matrix. If you denote the data (design or indicator matrix) as matrix X, then matrix product X'X is a Burt table).

Overall, the data matrix is symmetrical. In the case of 3 categorical variables, the data matrix consists 3 x 3 = 9 partitions, created by each variable being tabulated against itself, and against the categories of all other variables. Note that the sum of the diagonal elements in each diagonal partition (i.e., where the respective variables are tabulated against themselves) is constant (equal to 764 in this case). The off-diagonal elements in each partition in this example are all 0. If the cases in the design or indicator matrix are assigned to categories via fuzzy coding, then the off- diagonal elements of the diagonal partitions are not necessarily equal to 0.

C

C Named as one of a developmental sequence of theoretical programming languages : 'A', 'B' (also the useful language bcpl). A programming language of broad expressive power; thus suitable for both numerical and general programming. 'C' is closely associated with the construction of the ubiquitous computer operating system 'unix'. Compilers for 'C' are supplied for virtually all modern computers. 'C' is available as a standard programming language approved by ANSI and ISO.

C&RT C&RT, or Classification and Regression Trees, is a classification tree program developed by Breiman et. al. (1984).

Classification and Regression Trees are analytic procedure for predicting the values of a continuous response variable (e.g., Height) or categorical response variable (e.g., Marital Status: Single, Married, Divorced) from continuous or categorical predictors. When the dependent or response variable of interest is categorical in nature, the technique is referred to as Classification Trees; if the response variable of interest is continuous in nature, the method is referred to as Regression Trees. The classic computational algorithms for classification and regression trees (C&RT) were popularized by Breiman, Friedman, Olshen, & Stone, 1984. For classification problems, the goal is generally to find a tree where the terminal tree nodes are relatively "pure, " i.e., contain observations that (almost) all belong to the same category or class; for regression tree problems node purity is usually defined in terms of the sums-of-squares deviation within each node.

At each step, the program will find a logical split condition to assign observations to the two child nodes; for continuous predictors these logical conditions are usually of the type If x > Value then NodeID=k; for categorical predictors, the logical split conditions are usually of the type If x=Category I then NodeID=k.

Canonical correlation Canonical Correlation allows us to investigate the relationship between two sets of variables (it is used as either a hypothesis testing or exploratory method). For example, an educational researcher may want to compute the (simultaneous) relationship between three measures of scholastic ability with five measures of success in school. A sociologist may want to investigate the relationship between two predictors of social mobility based on interviews, with actual subsequent social mobility as measured by four different indicators. A medical researcher may want to study the relationship of various risk factors to the development of a group of symptoms. In all of these cases, the researcher is interested in the relationship between two sets of variables, and Canonical Correlation would be an appropriate method of analysis.

Capability ratio (C_r) This index is equivalent to C_p; specifically, it is computed as $1/C_p$ (the inverse of C_p).

Cartesian coordinates Cartesian coordinates (x, y, or x, y, z; also known as rectangular coordinates) are directed distances from two (or three) perpendicular axes.

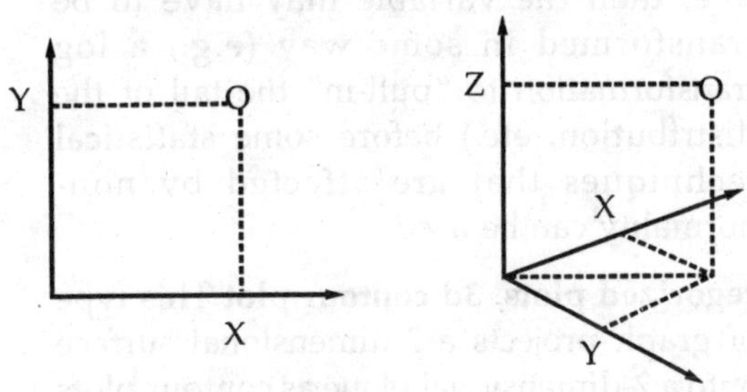

The location of a point in space is established by the corresponding coordinates on the X- and Y-axes (or X-, Y- , and Z-axes).

Case control study A retrospective study for examining factors contributing to a disease.

Casewise MD deletion When casewise deletion of missing data is selected, then only cases that do not contain any missing data for any of the variables selected for the analysis will be included in the analysis. In the case of correlations, all correlations are calculated by excluding cases that have missing data for any of the selected variables (all correlations are based on the same set of data).

Categorical data A set of data is said to be categorical if the values or observations belonging to it can be sorted according to category. Each value is chosen from a set of non-overlapping categories. For example, shoes in a cupboard can be sorted according to colour; the characteristic 'colour' can have non-overlapping categories 'black', 'brown', 'red' and 'other'. People have the characteristic of 'gender' with categories 'male' and 'female'.

Categories should be chosen carefully since a bad choice can prejudice the outcome of an investigation. Every value should belong to one and only one category, and there should be no doubt as to which one.

Categorical dependent variable A categorical dependent variable is a variable of interest (a researcher wants to predict), measured on a nominal scale, whose values identify class or group membership (e.g., Gender, with classes Male and Female; or Education, with classes No High School Degree, High School Degree, Some College, College Degree, Some Graduate School, Graduate Degree). A researcher may be interested in predicting the group membership of observations based on the values of some independent or predictor variables. For example, Credit Risk, with values Good and Bad, would be a categorical dependent variable that one might want to predict based on measured independent variables that are possibly related to credit risk.

Categorical predictor variable A categorical predictor variable is a variable, measured on a nominal scale, whose categories identify class or group membership, which is used to predict responses on one or more dependent variables. Gender would be an example of a categorical predictor variable, with the two classes or groups Male and Female.

Categorical variable Also called *qualitative* variable

A variable whose possible values are one of a number of categories. Although the values may be represented by numeric codes, the numbers are not thought of as measured or counted quantities. cf.quantitative variable.

Categorized contour/lines (ternary graph) This 3D Categorized Plot projects a 3-dimensional surface onto a 2-dimensional plane as line contour plots for each level of the grouping variable. One component graph is produced for each level of the grouping variable (or user-defined subset of data) and all the component graphs are arranged in one display to allow for comparisons between the subsets of data (categories).

Categorized graphs (also, trellis graphs) This type of graph allows you to categorize 2D, 3D, or nD plots by the specified categories of a selected variable. One component graph is produced for each level of the grouping variable (or user-defined subset of data) and all the component graphs are arranged in one display to allow for comparisons between the subsets of data (categories).

Categorized plots, 2d detrended probability plots This categorized plot is constructed in the same way as the standard normal probability plot for the categorized values, except that before the plot is generated, the linear trend is removed. This often "spreads out" the plot, thereby allowing the user to detect patterns of deviations more easily.

Categorized plots, 2d half-normal probability plots The categorized half-normal probability plot is constructed in the same manner as the standard normal probability plot, except that only the positive half of the normal curve is considered. Consequently, only positive normal values will be plotted on the Y-axis. This plot is often used in plots of residuals (e.g., in multiple regression), when one wants to ignore the sign of the residuals, that is, when one is interested in the distribution of absolute residuals, regardless of the sign.

Categorized plots, 2d normal probability plots This type of probability plot is constructed as follows. First, within each category, the values (observations) are rank ordered. From these ranks one can compute z values (i.e., standardized values of the normal distribution) based on the assumption that the data come from a normal distribution. These z values are plotted on the Y-axis in the plot. If the observed values (plotted on the X-axis) are normally distributed, then all values should fall onto a straight line. If the values are not normally distributed, then they will deviate from the line. Outliers may also become evident in this plot. If there is a general lack of fit, and the data seem to form a clear pattern (e.g., an S shape) around the line, then the variable may have to be transformed in some way (e.g., a log transformation to "pull-in" the tail of the distribution, etc.) before some statistical techniques that are affected by non-normality can be used.

Categorized plots, 3d contour plot This type of graph projects a 3-dimensional surface onto a 2-dimensional plane as contour plots for each level of the grouping variable. The plots are arranged in one display to allow for comparisons between the subsets of data (categories).

Categorized plots, 3d deviation plot Data points (representing the X, Y and Z coordinates of each point) in this graph are represented in 3D space as "deviations" from a specified base- level of the Z-axis. One component graph is produced for each level of the grouping variable (or user-defined subset of data) and all the component graphs are arranged in one display to allow for comparisons between the subsets of data (categories).

Categorized plots, 3d scatterplot This type of graphs visualizes a relationship between three variables (representing the X, Y, and one or more Z [vertical] coordinates of each point in 3- dimensional space) categorized by a grouping variable. One component graph is produced for each level of the grouping variable (or user-defined subset of data) and all the component graphs are arranged in one display to allow for comparisons between the subsets of data (categories).

Categorized plots, 3d space plot This type of graph offers a distinctive means of representing 3D scatterplot data through the use of a separate X-Y plane positioned at a user-selectable level of the vertical Z-axis (which "sticks up" through the middle of the plane). The level of the X-Y plane can be adjusted in order to divide the X-Y-Z- space into meaningful parts (e.g., featuring different patterns of the relation between the three variables).

Categorized plots, 3d surface plot In this type of graph , a surface (defined by a smoothing technique or user-defined mathematical expression) is fitted to the categorized data (variables corresponding to sets of XYZ coordinates, for subsets of data determined by the selected categorization method) arranged in one display to allow for comparisons between the subsets of data (categories).

Categorizing, grouping, slicing, drilling down One of the most important, general, and also powerful analytic methods involves dividing ("splitting") the data set into categories in order compare the patterns of data between the resulting subsets. This common technique is known under a variety of terms (such as breaking down, grouping, categorizing, splitting, slicing, drilling-down, or conditioning) and it is used both in exploratory data analyses and hypothesis testing. For example: A positive relation between the age and the risk of a heart attack may be different in males and females (it may be stronger in males). A promising relation between taking a drug and a decrease of the cholesterol level may be present only in women with a low blood pressure and only in their thirties and forties. The process capability indices or capability histograms can be different for periods of time supervised by different operators. The regression slopes can be different in different experimental groups.

There are many computational techniques that capitalize on grouping and that are designed to quantify the differences that the grouping will reveal (e.g., ANOVA/ MANOVA). However, graphical techniques (such as categorized graphs) offer unique advantages that cannot be substituted by any computational method alone: they can reveal patterns that cannot be easily quantified (e.g., complex interactions, exceptions, anomalies) and they provide unique, multidimensional, global analytic perspectives to explore or mine the data.

Cauchy distribution The Cauchy distribution (the term first used by Upensky, 1937) has density function:

$f(x) = 1/(\theta\pi^*\{1 + [(x-\eta)/\theta]^2\})$

$0 < \theta$

where η is the location parameter (median)

θ is the scale parameter

is the constant Pi (3.14...).'

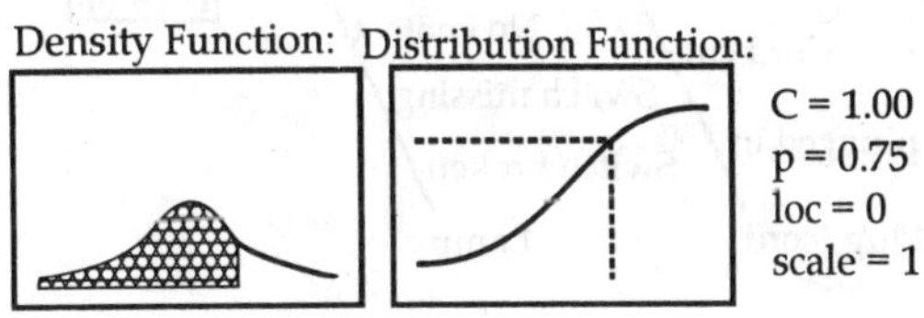

Causation, causal relation Two variables are causally related if changes in the value of one cause the other to change. For example, if one heats a rigid container filled with a gas, that causes the pressure of the gas in the container to increase. Two variables can be associated without having any causal relation, and even if two variables have a causal relation, their correlation can be small or zero.

Cause and effect diagram The Cause-and-Effect diagram provides an efficient summary of factors that impact a process, and hence can be used as a map to guide the overall quality improvement efforts. Therefore, it is one of the important tools for the Define phase of Six Sigma quality control efforts. The diagram is also sometimes referred to as a "fishbone chart," because of it's appearance, or an Ishikawa chart. The latter name refers to the work of Professor Kaoru Ishikawa of Tokyo University who developed this diagram to depict variables which are present in a process. The general idea of the chart is rather straightforward. Suppose you want to turn on a reading light in your house one evening, and it won't light up. Now consider the various variables or characteristics that make up the process (cause the light to come on), and which should be considered in order to fix this quality problem.

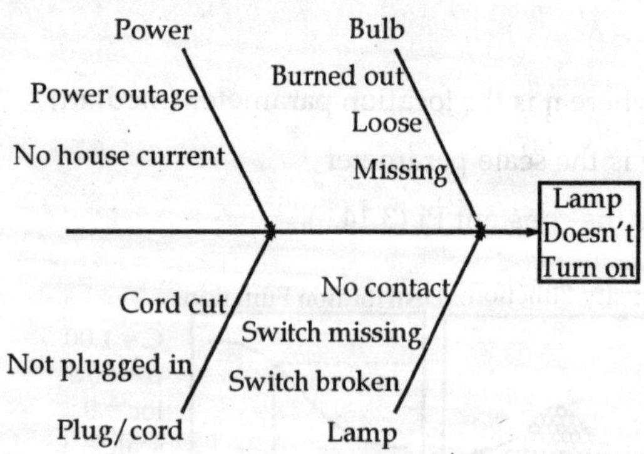

Usually, the chart is constructed by identifying 1. The major categories of causes that affect the process (in this example Power, Bulb, Plug/Cord and Lamp), and

2. The individual factors or causes that can be classified into these major categories (e.g., Power outage, No house current, etc.). You could now use this map as a guide to troubleshooting the problem you encountered turning on your reading light. You can also further "augment" this chart by adding various sub-sub causes, causes that you ruled out, solutions you have tried, etc.

The cause-and-effect diagram plays a central role in Six Sigma quality programs. During the first stage of the define-measure-analyze-improve-control (DMAIC) cycle, this diagram can be of great utility in order to identify the areas, departments, processes, and stakeholders that should be involved in the effort.

CDF See cumulative distribution function

Cell The 'hole' into which a number (a frequency, proportion or percentage) is put when constructing a table.

Censored Referring to a variable about whose value we might only know a lower or upper bound. Variables representing survival time are typically right censored for individuals still alive at the time data are collected, or for individuals who die from a cause other than the one being studied. Measurements of concentrations that may be below the detection limit of the method employed are an example of left censored variables.

Censoring (censored observations) Observations are referred to as censored when the dependent variable of interest represents the time to a terminal event, and the duration of the study is limited in time. Although the concept was developed in the biomedical research, censored observations may occur in a number of different areas of research. For example, in the social sciences we may study the "survival" of marriages, high school drop-out rates (time to drop-out), turnover in organizations, etc. In each case, by the end of the study period, some subjects probably will still be married, will not have dropped out, or will still be working at the same company; thus, those subjects represent censored observations.

In economics we may study the "survival" of new businesses or the "survival" times of products such as automobiles. In quality control research, it is common practice to study the "survival" of parts under stress (failure time analysis).

Data sets with censored observations can be analyzed via Survival Analysis or via Weibull and Reliability/Failure Time Analysis.

Census A survey of every case in a population

Central limit theorem The central limit theorem states that the probability histograms of the sample mean and sample sum of n draws with replacement from a box of labeled tickets converge to a normal curve as the sample size n grows, in the following sense: As n grows, the area of the probability histogram for any range of values approaches the area under the normal curve for the same range of values, converted to standard units.

Centroid The point on a scatterplot with coordinates given by the mean values of the explanatory and response variables.

Certain event An event is *certain* if its probability is 100%. Even if an event is certain, it might not occur. However, by the complement rule, the chance that it does not occur is 0%.

Chaid Chaid is a classification trees program developed by Kass (1980) that performs multi-level splits when computing classification trees.

Chance variation, chance error A random variable can be decomposed into a sum of its expected value and chance variation around its expected value. The expected value of the chance variation is zero; the standard error of the chance variation is the same as the standard error of the random variable—the size of a "typical" difference between the random variable and its expected value.

Change of units or variables See also transformation.

Characteristic life In Weibull and Reliability/Failure Time Analysis the characteristic life is defined as the point in time where 63.2 percent of the population will have failed; this point is also equal to the respective scale parameter b of the two-parameter Weibull distribution (with $\theta = 0$; otherwise it is equal to $b+\theta$).

Chebychev's inequality For lists: For every number $k>0$, the fraction of elements in a list that are k SD's or further from the arithmetic mean of the list is at most $1/k^2$. For random variables: For every number $k>0$, the probability that a random variable X is k SEs or further from its expected value is at most $1/k^2$.

Chi square A probability distribution used to test the independence of two nominal variables.

Chi square curve The chi-square curve is a family of curves that depend on a parameter called degrees of freedom (*d.f.*). The chi-square curve is an approximation to the probability histogram of the *chi-square statistic* for multinomial model if the expected number of outcomes in each category is large. The chi-square curve is positive, and its total area is 100%, so we can think of it as the probability histogram of a random variable. The balance point of the curve is *d.f.*, so the expected value of the corresponding random variable would equal *d.f.*. The standard error of the corresponding random variable would be $(2\times d.f.)^{1/2}$. As *d.f.* grows, the shape of the chi-square curve approaches the shape of the normal curve.

Chi square distribution The Chi-square distribution is defined by:

$$f(x) = \{1/[2^{\nu/2} * \Gamma(\nu/2)]\} * [x^{(\nu/2)-1} * e^{-x/2}]$$

$\nu = 1, 2, ..., 0 < x$

where ν is the degrees of freedom

e is the base of the natural logarithm, sometimes called Euler's e

Γ (gamma) is the Gamma function.

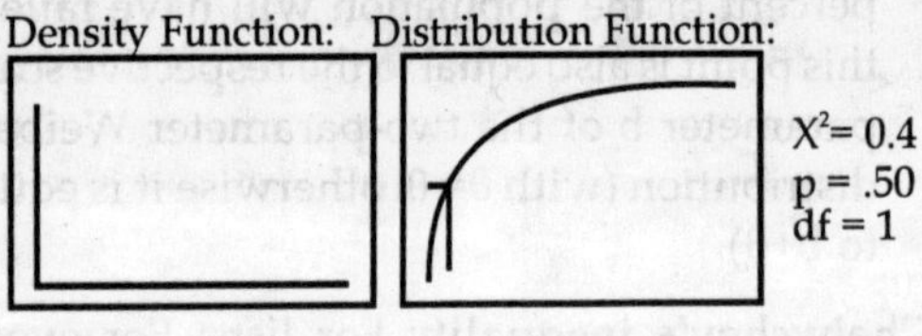

Chi square statistic The *chi-square* statistic is used to measure the agreement between categorical data and a multinomial model that predicts the relative frequency of outcomes in each possible category. Suppose there are *n* independent trials, each of which can result in one of *k* possible outcomes. Suppose that in each trial, the probability that outcome *i* occurs is p_i, for $i = 1, 2, \ldots, k$, and that these probabilities are the same in every trial. The expected number of times outcome 1 occurs in the *n* trials is $n \times p_1$; more generally, the expected number of times outcome *i* occurs is

$\text{expected}_i = n \times p_i$.

If the model be correct, we would expect the *n* trials to result in outcome *i* about $n \times p_i$ times, give or take a bit. Let observed_i denote the number of times an outcome of type *i* occurs in the *n* trials, for $i = 1, 2, \ldots, k$. The *chi-squared statistic* summarizes the discrepancies between the expected number of times each outcome occurs (assuming that the model is true) and the observed number of times each outcome occurs, by summing the squares of the discrepancies, normalized by the expected numbers, over all the categories:

chi-squared =

$$(\text{observed}_1 - \text{expected}_1)^2/\text{expected}_1 + (\text{observed}_2 - \text{expected}_2)^2/\text{expected}_2 + \ldots + (\text{observed}_k - \text{expected}_k)^2/\text{expected}_k.$$

As the sample size *n* increases, if the model is correct, the sampling distribution of the *chi-squared statistic* is approximated increasingly well by the chi-squared curve with

(#categories – 1) = *k* – 1

degrees of freedom (*d.f.*), in the sense that the chance that the *chi-squared statistic* is in any given range grows closer and closer to the area under the Chi-Squared curve over the same range.

Chi squared goodness of fit test The Chi-Squared Goodness of Fit Test is a test for comparing a theoretical distribution, such as a Normal, Poisson etc, with the observed data from a sample.

Chi squared test of association The Chi-Squared Test of Association allows the comparison of two attributes in a sample of data to determine if there is any relationship between them.

The idea behind this test is to compare the observed frequencies with the frequencies that would be expected if the null hypothesis of no association / statistical independence were true. By assuming the variables are independent, we can also predict an expected frequency for each cell in the contingency table.

If the value of the test statistic for the chi-squared test of association is too large, it indicates a poor agreement between the observed and expected frequencies and the null hypothesis of independence / no association is rejected.

Chi squared test of homogeneity On occasion it might happen that there are several proportions in a sample of data to be tested simultaneously. An even more complex situation arises when the several populations have all been classified according to the same variable. We generally do not expect an equality of proportions for all the classes of all the populations. We do however, quite often need to test whether the proportions for each class are equal across all populations and whether this is true for each class. If this proves to be the case, we say the populations are homogeneous with respect to the variable of classification. The test used for this purpose is the Chi-Squared Test of Homogeneity, with hypotheses,

H0: the populations are homogeneous with respect to the variable of classification against

H1: the populations are not homogeneous

Circumplex A circumplex is a set of variables which, when plotted as vectors in N-dimensional space, fall in a circular pattern. If a set of variables forms a circumplex, the correlation matrix for the variables will have an unusual pattern called circular structure. In this pattern, the correlations on diagonal strips below the main diagonal tend to be equal, or nearly so, first becoming smaller, then larger again as you move away from the main diagonal.

Circumplex is a special case of a more general concept of radex, developed by Louis Guttman (who contributed a number of innovative ideas to the theory of multidimensional scaling and factor analysis, Guttman, 1954).

City block The error is the sum of the differences between the target and actual output values on each output unit; differences are always taken to be positive. The city-block error function is less sensitive to outlying points than the sum-squared error function (where a disproportionate amount of the error can be accounted for by the worst-behaved cases). Consequently, networks trained with this metric may perform better on regression problems if there are a few wide-flung outliers (either because the data naturally has such a structure, or because some cases may be mislabeled).

City block (manhattan) distance A distance measure computed as the average difference across dimensions. In most cases, this distance measure yields results similar to the simple Euclidean distance. However, note that in this measure, the effect of single large differences (outliers) is dampened (since they are not squared). See, Cluster Analysis.

City block error function Defines the error between two vectors as the sum of the differences in each component. Less sensitive to outliers than sum-squared, but usually causes poorer training performance.

Class boundary A point that is the left endpoint of one class interval, and the right endpoint of another class interval.

Class frequency The number of observations that fall into each class interval.

Class interval In plotting a histogram, one starts by dividing the range of values into a set of non-overlapping intervals, called *class intervals,* in such a way that every datum is contained in some class interval.

Class intervals Categories or groups contained in frequency graphics.

Class labeling A variety of clustering algorithms and networks are supported. All of these have networks where the second layer consists of Radial units, and these units contain exemplar vectors. In SOFM and Cluster networks, these are combined in a two layer neural network with a single nominal output variable and the KNearest output conversion function to produce a classification based upon the nearest exemplar vector(s) to an input case.

The exemplar vectors can be positioned using a variety of cluster-center and sampling approaches. However, it is also necessary to apply class labels to the radial units (i.e. label each radial unit as representative of a particular class). Class labels can be applied using the class labeling algorithms described here.

Classification In classification, the aim is to assign input cases to one of a number of classes.

Classification problems fall into two categories: two-class problems, and many-class problems. A two-class problem is usually encoded using a single output neuron. Many-class problems use one output neuron per class. It is also possible to encode two-class problems using this approach (i.e., using two output neurons), and this is in fact the approach taken by PNN networks.

Classification by labeled exemplars Classification neural networks must translate the numeric level on the output neuron(s) to a nominal output variable. There are two very different approaches to assigning classifications.

In one of these, the activation level of the output layer units determines the class, usually by interpreting the activation level as a confidence measure, and finding the highest confidence class. That approach is used in most neural network types.

This topic discusses the alternative approach, which is used in SOFM and Cluster networks.

These types of networks store labeled exemplar vectors in their radial layer. When a new case is presented to the network, the network in essence calculates the distance between the (possibly normalized) new case and each exemplar vector; the activations of the neurons encode these distances. Each of these neurons has a class label. The class label of the "winning" (smallest distance from input case) neuron is typically used as the output of the network. The standard algorithm is extended slightly using the KL nearest neighbor algorithm; the class assigned by the network is the most common class among the K winning neurons, provided that at least L of them agree (otherwise, the class is "Unknown").

The input case might actually be very distant from any of the exemplar vectors, in which case it may be better to assign the case as "Unknown." You can optionally specify an accept threshold for this eventuality. If the normalized distance is greater than this threshold, the class is "Unknown."

Classification statistics One of the major uses of neural networks is to perform classification tasks, i.e., to assign cases to one of a number of possible classes. The class of a case is indicated by the use of a nominal output variable.

Classification thresholds Classification neural networks must translate the numeric activation level of the output neuron(s) to a nominal output variable. There are two very different approaches to assigning classifications.

One of these, where the neural network determines the "winning" neuron or neurons in the radial layer of the network, and then uses the class labels on those neurons, is described in Classification by Labeled Exemplars. This approach is used in SOFM and Cluster networks.

Here we discuss the alternative approach, where it is the activation level of the output layer units, which are not radial units, that determines the class. This approach is used in all other network types.

Two cases need to be distinguished: single output neuron versus multiple output neurons.

Single output neurons are typically used for two-class problems, with a high output neuron level indicating one class and a low activation the other class. This configuration uses the Two-state conversion function.

Multiple output neurons are typically used for three or more class problems. One neuron is used for each class, and the highest activation neuron indicates the class. The neuron activation levels can be interpreted as confidence levels. This method is implemented by using the One-of-N conversion function for the output variable. You can optionally configure a multilayer perceptron to use two output neurons for a two-class output variable.

Classification trees Classification trees are used to predict membership of cases or objects in the classes of a categorical dependent variable from their measurements on one or more predictor variables.

Cluster analysis The term cluster analysis (first used by Tryon, 1939) actually encompasses a number of different classification algorithms which can be used to develop taxonomies (typically as part of exploratory data analysis). For example, biologists have to organize the different species of animals before a meaningful description of the differences between animals is possible. According to the modern system employed in biology, man belongs to the primates, the mammals, the amniotes, the vertebrates, and the animals. Note how in this classification, the higher the level of aggregation the less similar are the members in the respective class. Man has more in common with all other primates (e.g., apes) than it does with the more "distant" members of the mammals (e.g., dogs), etc.

Cluster diagram A scatter diagram plotting cases belonging to various classes in two dimensions. The dimensions are provided by the output levels of units in the neural network. See also, Cluster Analysis.

Cluster networks Cluster networks are actually a non-neural model presented in a neural form for convenience. A cluster network consists of a number of class-labeled exemplar vectors (each represented by a radial neuron). The vectors are assigned centers by clustering algorithms such as K-Means, and then labeled using nearby cases. After labeling, the centers positions can be fine-tuned using Learned Vector Quantization.

Cluster networks are closely related to Kohonen networks, with a few differences. Cluster networks are intended for supervised learning situations where class labels are available in the training data and cluster networks do not have a topologically organized output layer.

Training consists of center assignment, followed by labeling. This can optionally be followed by LVQ training to improve center location. After training, the network can be pruned, and classification factors set, including an acceptance threshold and K,L factors for KL nearest neighbor classification.

Cluster sample In a cluster sample, the sampling unit is a collection of population units, not single population units. For example, techniques for adjusting the U.S. census start with a sample of geographic blocks, then (try to) enumerate all inhabitants of the blocks in the sample to obtain a sample of people. This is an example of a cluster sample. (The blocks are chosen separately from different strata, so the overall design is a stratified cluster sample.)

Cluster sampling Cluster sampling is a sampling technique where the entire population is divided into groups, or clusters, and a random sample of these clusters are selected. All observations in the selected clusters are included in the sample.

Cluster sampling is typically used when the researcher cannot get a complete list of the members of a population they wish to study but can get a complete list of groups or 'clusters' of the population. It is also used when a random sample would produce a list of subjects so widely scattered that surveying them would prove to be far too expensive, for example, people who live in different postal districts in the U.K.

This sampling technique may well be more practical and/or economical than simple random sampling or stratified sampling.

Example

Suppose that the Department of Agriculture wishes to investigate the use of pesticides by farmers in England. A cluster sample could be taken by identifying the different counties in England as clusters. A sample of these counties (clusters) would then be chosen at random, so all farmers in those counties selected would be included in the sample. It can be seen here then that it is easier to visit several farmers in the same county than it is to travel to each farm in a random sample to observe the use of pesticides.

Codebook A document that lists the dictionary information about all the variables in a data file. This usually includes the original question text, the spss variable names if appropriate, and the value labels for each coded response. A codebook may also contain information about derived variables and notes given to interviewers or coders when preparing the data file.

Codes Codes are values of a grouping variable (e.g., 1, 2, 3, ... or MALE, FEMALE) which identify the levels of the grouping variable in an analysis. Codes can either be text values or integer values.

Coding The process by which numbers are ascribed to responses to a survey questionnaire in preparation for computer analysis. For example, to the question "Are you male or female?", male may be coded with the number 1 and female with the number 2.

Coefficient of correlation A number that represents the direction and strength of a correlation.

Coefficient of determination The percentage of variation accounted for in one variable by knowing the value of another variable.

Coefficient of variation The coefficient of variation measures the spread of a set of data as a proportion of its mean. It is often expressed as a percentage.

It is the ratio of the sample standard deviation to the sample mean:

$$\frac{s}{\bar{x}}$$

There is an equivalent definition for the coefficient of variation of a population, which is based on the expected value and the standard deviation of a random variable:

$$\frac{\sqrt{V(X)}}{E(X)}$$

Cohort A collection of individuals of the same age followed through time.

Column percentage A percentage calculated by finding the count in a cell compared with the total count for the column (the column marginal count). For example, if there are 530 unemployed women in a sample including 1192 women in all, the column percentage of unemployed women is 530/1192 %, or 45%.

Columns (box plot) In this type of box plot, vertical columns are used to represents the variable's midpoint (i.e., mean or median).

The whiskers superimposed on each column mark the selected range (i.e., standard error, standard deviation, min-max, or constant) around the midpoint.

Combinations The number of combinations of n things taken k at a time is the number of ways of picking a subset of k of the n things, without replacement, and without regard to the order in which the elements of the subset are picked. The number of such combinations is ${}_nC_k = n!/(k!(n-k)!)$, where $k!$ (pronounced "k factorial") is $k\times(k-1)\times(k-2)\times \cdots \times 1$. The numbers ${}_nC_k$ are also called the *Binomial coefficients*. From a set that has n elements one can form a total of 2^n subsets of all sizes. For example, from the set {a, b, c}, which has 3 elements, one can form the $2^3 = 8$ subsets {}, {a}, {b}, {c}, {a,b}, {a,c}, {b,c}, {a,b,c}. Because the number of subsets with k elements one can form from a set with n elements is ${}_nC_k$, and the total number of subsets of a set is the sum of the numbers of possible subsets of each size, it follows that ${}_nC_0+{}_nC_1+{}_nC_2+ \ldots +{}_nC_n = 2^n$. The calculator has a button (nCm) that lets you compute the number of combinations of m things chosen from a set of n things. To use the button, first type the value of n, then push the nCm button, then type the value of m, then click the "=" button.

Communality In Principal Components and Factor Analysis, Communality is the proportion of variance that each item has in common with other items. The proportion of variance that is unique to each item is then the respective item's total variance minus the communality. A common starting point is to use the squared multiple correlation of an item with all other items as an estimate of the communality (refer to Multiple Regression for details about multiple regression). Some authors have suggested various iterative "post- solution improvements" to the initial multiple regression communality estimate; for example, the so-called MINRES method (minimum residual factor method; Harman & Jones, 1966) will try various modifications to the factor loadings with the goal to minimize the residual (unexplained) sums of squares.

Compiler A program supplied especially for a particular type of computer, to enable the translation of code expressed in some programming language into object code for that computer. A compiler undertakes translation of the whole of the user's program to produce an object code version which is complete,

Complement The complement of a subset of a given set is the collection of all elements of the set that are not elements of the subset.

Complement rule The probability of the complement of an event is 100% minus the probability of the event: $P(A^c) = 100\% - P(A)$.

Completely randomised design The structure of the experiment in a completely randomised design is assumed to be such that the treatments are allocated to the experimental units completely at random.

See also treatment.

See also experimental unit.

Complex numbers Complex numbers are the superset that includes all real and imaginary numbers. A complex number is usually represented by the expression a + ib where a and b are real numbers and i is the imaginary part of the expression where i has the property that $i^2=-1$.

See also, Cross-spectrum Analysis.

Composite hypothesis A composite hypothesis is a hypothesis which does not specify the population distribution completely.

Examples

1. X~Bi(100,p) H1: $p > 0.5$
2. X~N(0, σ^2) H1: σ^2 unspecified.

Computer An automatic data-processing device which is programmable.

Computer program A specification of how to undertake a certain process, usually expressed via a programming language, for some chosen computer. Also see : program.

Concentration ellipse An ellipse that encompasses the majority of points in a scatter diagram.

Concordant If there are two cases, A and B, and case A has a higher value on one variable than case B, and case A also has a higher value on another variable than case B, the two cases are concordant for the two variables.

Condition An event which is assumed to have occurred when considering the probability of another event.

Conditional probability Suppose we are interested in the probability that some event A occurs, and we learn that the event B occurred. How should we update the probability of A to reflect this new knowledge? This is what the conditional probability does: it says how the additional knowledge that B occurred should affect the probability that A occurred quantitatively. For example, suppose that A and B are mutually exclusive. Then if B occurred, A did not, so the *conditional probability that A occurred given that B occurred* is zero. At the other extreme, suppose that B is a subset of A, so that A must occur whenever B does. Then if we learn that B occurred, A must have occurred too, so the *conditional probability that A occurred given that B occurred* is 100%. For in-between cases, where A and B intersect, but B is not a subset of A, the conditional probability of A given B is a number between zero and 100%. Basically, one "restricts" the outcome space S to consider only the part of S that is in B, because we know that B occurred. For A to have happened given that B happened requires that AB happened, so we are interested in the event AB. To have a legitimate probability requires that P(S) = 100%, so if we are restricting the outcome space to B, we need to divide by the probability of B to make the probability of this new S be 100%. On this scale, the probability that AB happened is P(AB)/P(B). This is the definition of the conditional probability of A given B, provided P(B) is not zero (division by zero is undefined). Note that the special cases AB = {} (A and B are mutually exclusive) and AB = B (B is a subset of A) agree with our intuition as described at the top of this paragraph. Conditional probabilities satisfy the axioms of probability, just as ordinary probabilities do.

Confidence ensembles Confidence ensembles are much more restrictive than output ensembles. The network predictions are combined at the level of the output neurons. To make sense, the encoding of the output variables must therefore be the same for all the members. Given that restriction, there is no point in forming confidence ensembles for regression problems, as the effect is to produce the same output as an output ensemble, but with the averaging performed before scaling rather than after. Confidence ensembles are designed for use with classification problems.

The advantage of using a confidence ensemble for a classification problem is that it can estimate overall confidence levels for the various classes, rather than simply providing a final choice of class.

Confidence interval For a given re-randomisation distribution, a family of

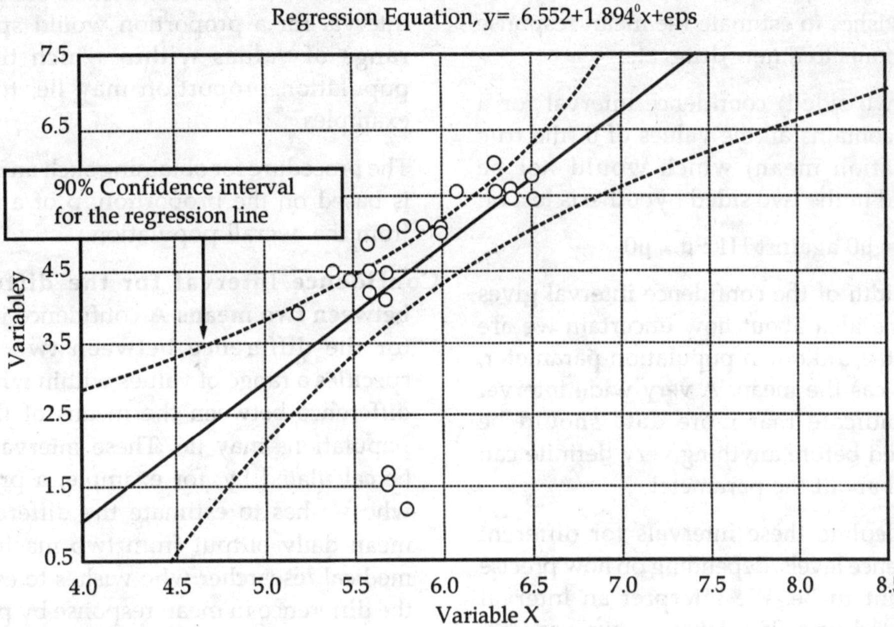

related distributions may be defined according to a range of hypothetical values of the pattern which the test statistic measures. For instance, for the pitman permutation test(2) to test for a scale shift between two groups, a related distribution may. be formed by shifting all the observations in one group by a common amount, where this common shift is regarded as a continuous variable. With finite numbers of data the number of related distributions will be finite, and typically considerably smaller than the number of points of the randomisation distribution. The likelihood of the outcome value may be calculated for each distribution in the family, and these likelihoods may be then used to define a contiguous set of values which occupy a certain proportion of the total unit weight of the likelihoods integrated over all values of the test statistic. The confidence interval is defined by the minimum and maximum values of the range of values so defined. The proportion of the total weight within the range of values is regarded as an alpha probability that the value of the test statistic lies within this range. Generally the definition of a confidence interval cannot be unique without imposing further constraints. Approaches to providing suitable constraints, such that a confidence interval will be unique, include defining the confidence interval : to include the whole of one tail of the distribution; or to be centred in some sense upon the outcome value; or to be centred between tails of equal weight. In the case of re-randomisation distributions, these are discrete distributions so there will generally be no range of values with weight corresponding exactly to an arbitrary nominal alpha criterion level, and the problem of non-uniqueness is therefore not generally solvable.

Confidence interval for a mean A confidence interval for a mean specifies a range of values within which the unknown population parameter, in this case the mean, may lie.

These intervals may be calculated by, for example, a producer who wishes to estimate his mean daily output; a medical researcher who wishes to estimate the mean response by patients to a new drug; etc.

The (two sided) confidence interval for a mean contains all the values of 0 (the true population mean) which would not be rejected in the two sided hypothesis test of:

H0 : $\mu = \mu0$ against H1 : $\mu \neq \mu0$

The width of the confidence interval gives us some idea about how uncertain we are about the unknown population parameter, in this cas the mean. A very wide interval may indicate that more data should be collected before anything very definite can be said about the parameter.

We calculate these intervals for different confidence levels, depending on how precise we want to be. We interpret an interval calculated at a 95% level as, we are 95% confident that the interval contains the true population mean. We could also say that 95% of all confidence intervals formed in this manner (from different samples of the population) will include the true population mean.

Compare one sample t-test.

Confidence interval for a proportion A confidence interval gives us some idea of the range of values which an unknown population parameter (such as the mean μ, variance σ^2) is likely to take based on a given set of sample data.

Sometimes we are interested in the proportion of responses that fall into one of two categories. For example, a firm may wish to know what proportion of their customers pay by credit card as opposed to those who pay by cash; the manager of a T.V. station may wish to know what percentage of households in a certain town have more than one T.V. set; a doctor may be interested in the proportion of patients who benefited from a new drug as opposed to those who didn't, etc. A confidence interval for a proportion would specify a range of values within which the true population proportion may lie, for such examples.

The procedure for obtaining such an interval is based on the proportion, p of a sample from the overall population.

Confidence interval for the difference between two means A confidence interval for the difference between two means specifies a range of values within which the difference between the means of the two populations may lie. These intervals may be calculated by, for example, a producer who wishes to estimate the difference in mean daily output from two machines; a medical researcher who wishes to estimate the difference in mean response by patients who are receiving two different drugs; etc.

The confidence interval for the difference between two means contains all the values of:

$\mu_1 - \mu_2$

(the difference between the two population means) which would not be rejected in the two sided hypothesis test of:

H0: $\mu_1 = \mu_2$ against H1: $\mu_1 \neq \mu_2$ i.e.

H0: $\mu_1 - \mu_2 = 0$ against H1: $\mu_1 - \mu_2 \neq 0$.

If the confidence interval includes 0 we can say that there is no significant difference between the means of the two populations, at a given level of confidence.

The width of the confidence interval gives us some idea about how uncertain we are about the difference in the means. A very wide interval may indicate that more data should be collected before anything definite can be said.

We calculate these intervals for different confidence levels, depending on how precise we want to be. We interpret an interval calculated at a 95% level as, we are 95% confident that the interval contains the true difference between the two population means. We could also say that 95% of all confidence intervals formed in this manner (from different samples of the population) will include the true difference.

Confidence interval for the difference between two proportions A confidence interval gives us some idea of the range of values which an unknown population parameter (such as the mean μ, variance σ^2) is likely to take based on a given set of sample data.

Many occasions arise where we have to compare the proportions of two different populations. For example, a firm may want to compare the proportions of defective items produced by different machines; medical researchers may want to compare the proportions of men and women who suffer heart attacks etc. A confidence interval for the difference between two proportions would specify a range of values within which the difference between the two true population proportions may lie, for such examples.

The procedure for obtaining such an interval is based on the sample proportions, p1 and p2, from their respective overall populations.

Confidence interval vs. prediction interval In regression, it is possible to predict the value of the dependent variable based on given values of the independent variables. When these values are predicted, it is also possible to calculate confidence intervals and/or prediction intervals for the dependent variable.

The confidence interval gives information on the expected value (mean) of the dependent variable. That is, a confidence interval for a predicted value of the dependent variable gives a range of values around which the "true" (population) mean (of the dependent variable for given levels of the independent variables) can be expected to be located.

The prediction interval gives information on individual predictions of the dependent variable. That is, a prediction interval for a predicted value of the dependent variable gives us a range of values around which an additional observation of the dependent variable can be expected to be located (with a given level of certainty, see also Elementary Concepts).

Note that the confidence interval will produce a smaller range of values, because it is an interval estimate for an average rather than an interval estimate for a single observation.

Confidence level The confidence level is the probability value $(1 - \alpha)$ associated with a confidence interval.

It is often expressed as a percentage. For example, say $\alpha = 0.05 = 5\%$, then confidence level = (1-0.05) = 0.95, that is, a 95% confidence level.

Example

Suppose an opinion poll predicted that, if the election were held today, the Conservative party would win 60% of the vote. The pollster might attach a 95% confidence level to the interval 60% plus or minus 3%. That is, he thinks it very likely that the Conservative party would get between 57% and 63% of the total vote.

Confidence limits Confidence limits are the lower and upper boundaries / values of a confidence interval, that is, the values which define the range of a confidence interval.

The upper and lower bounds of a 95% confidence interval are the 95% confidence limits. These limits may be taken for other confidence levels, for example, 90%, 99%, 99.9%

Confidence value (association rules) When applying (in data or text mining) algorithms for deriving association rules of the general form If Body then Head (e.g., If (Car=Porsche and Age<20) then (Risk=High and Insurance=High)), the Confidence value denotes the conditional probability of the Head of the association rule, given the Body of the association rule.

Confirmatory analysis A statistical analysis that aims to test a pre-specified hypothesis or model.

Confounding When the differences between the treatment and control groups other than the treatment produce differences in response that are not distinguishable from the effect of the treatment, those differences between the groups are said to be *confounded* with the effect of the treatment (if any). For example, prominent statisticians questioned whether differences between individuals that led some to smoke and others not to (rather than the act of smoking itself) were responsible for the observed difference in the frequencies with which smokers and non-smokers contract various illnesses. If that were the case, those factors would be confounded with the effect of smoking. Confounding is quite likely to affect observational studies and experiments that are not randomized. Confounding tends to be decreased by randomization. See also Simpson's Paradox.

Confusion matrix A name sometimes given to a matrix, in a classification problem, displaying the numbers of cases actually belonging to each class, and assigned by the neural network to that or other classes. Displayed as classification statistics in statistical neural networks.

Conjugate gradient descent Conjugate gradient descent is an advanced method of training multilayer perceptrons. It usually performs significantly better than back propagation, and can be used wherever back propagation can be. It is the recommended technique for any network with a large number of weights (more than a few hundred) and/or multiple output units. For smaller networks, either Quasi-Newton or Levenberg-Marquardt may be better, the latter being preferred for low-residual regression problems.

Conjugate gradient descent is a batch update algorithm: whereas back propagation adjusts the network weights after each case, conjugate gradient descent works out the average gradient of the error surface across all cases before updating the weights once at the end of the epoch.

For this reason, there is no shuffle option available with conjugate gradient descent, since it would clearly serve no useful function. There is also no need to select learning or momentum rates for conjugate gradient descent, so it can be much easier to use than back propagation. Additive noise would destroy the assumptions made by conjugate gradient descent about the shape of search space, and so is also not available.

Conjugate gradient descent works by constructing a series of line searches across the error surface. It first works out the direction of steepest descent, just as back propagation would do. However, instead of taking a step proportional to a learning rate, conjugate gradient descent projects a straight line in that direction and then locates a minimum along this line, a process that is quite fast as it only involves searching in one dimension. Subsequently, further line searches are conducted (one per epoch). The directions of the line searches (the conjugate directions) are chosen to try to ensure that the directions that have already been minimized stay minimized (contrary to intuition, this does not mean following the line of steepest descent each time).

The conjugate directions are actually calculated on the assumption that the error surface is quadratic, which is not generally the case. However, it is a fair working assumption, and if the algorithm discovers that the current line search direction isn't actually downhill, it simply calculates the line of steepest descent and restarts the search in that direction. Once a point close to a minimum is found, the quadratic assumption holds true and the minimum can be located very quickly.

Note The line searches on each epoch of conjugate gradient descent actually involve one gradient calculation plus a variable number (perhaps as high as twenty) of error evaluations. Thus a conjugate gradient descent epoch is substantially more time-consuming (typically 3-10 times longer) than a back propagation epoch. If you want to compare performance of the two algorithms, you will need to record the time taken.

Contingency table A contingency table is a way of summarising the relationship between variables, each of which can take only a small number of values. It is a table of frequencies classified according to the values of the variables in question.

When a population is classified according to two variables it is said to have been 'cross-classified' or subjected to a two-way classification. Higher classifications are also possible.

A contingency table is used to summarise categorical data. It may be enhanced by including the percentages that fall into each category.

What you find in the rows of a contingency table is contingent upon (dependent upon) what you find in the columns.

Continuity correction In using the normal approximation to the binomial probability histogram, one can get more accurate answers by finding the area under the normal curve corresponding to half-integers, transformed to standard units. This is clearest if we are seeking the chance of a particular number of successes. For example, suppose we seek to approximate the chance of 10 successes in 25 independent trials, each with probability p = 40% of success. The number of successes in this scenario has a binomial distribution with parameters n = 25 and p = 40%. The expected number of successes is $np = 10$, and the standard error is $(np(1-p))^{1/2} = 6^{1/2} = 2.45$. If we consider the area under the normal curve at the point 10 successes, transformed to standard units, we get zero: the area under a point is always zero. We get a better approximation by considering 10 successes to be the range from 9 1/2 to 10 1/2 successes. The only possible number of successes between 9 1/2 and 10 1/2 is 10, so this is exactly right for the binomial distribution. Because the normal curve is continuous and a binomial random variable is discrete, we need to "smear out" the binomial probability over an appropriate range. The lower endpoint of the range, 9 1/2 successes, is (9.5 - 10)/2.45 = -0.20 standard units. The upper endpoint of the range, 10 1/2 successes, is (10.5 – 10)/2.45 = +0.20 standard units. The area under the normal curve between -0.20 and +0.20 is about 15.8%. The true binomial probability is ${}_{25}C_{10}\times(0.4)^{10}\times(0.6)^{15}$ = 16%. In a similar way, if we seek the normal approximation to the probability that a binomial random variable is in the range from i successes to k successes, inclusive, we should find the area under the normal curve from i-1/2 to k+1/2 successes, transformed to standard units. If we seek the probability of more than i successes and fewer than k successes, we should find the area under the normal curve corresponding to the range i+1/2 to k-1/2 successes, transformed to standard units. If we seek the probability of more than *i* but no more than *k* successes, we should find the area under the normal curve corresponding to the range *i*+1/2 to *k*+1/2 successes, transformed to standard units. If we seek the probability of at least *i* but fewer than *k* successes, we should find the area under the normal curve corresponding to the range *i*-1/2 to *k*-1/2 successes, transformed to standard units. Including or excluding the half-integer ranges at the ends of the interval in this manner is called the continuity correction.

Continuous data A set of data is said to be continuous if the values / observations belonging to it may take on any value within a finite or infinite interval. You can count, order and measure continuous data. For example, height; weight; temperature; the amount of sugar in an orange; the time required to run a mile.

Compare discrete data.

Continuous dependent variable A continuous dependent variable is a variable of interest (a researcher wants to predict) measured on a continuous scale (such as Weight in pounds, or Height in centimeters). A researcher may be interested in predicting a continuous variable (e.g., Annual Sales) based on one or more predictors (e.g., based on Marketing Expenditures).

Continuous distribution A probability distribution of a continuous statistic, based upon an algebraic formula, such that for any possible value of the cumulative probability there is an exact corresponding value of the statistic in question. Also see : discrete distribution.

Continuous random variable A continuous random variable is one which takes an infinite number of possible values. Continuous random variables are usually measurements. Examples include height, weight, the amount of sugar in an orange, the time required to run a mile.

Compare discrete random variable.

Continuous variable A quantitative variable is *continuous* if its set of possible values is uncountable. Examples include temperature, exact height, exact age (including parts of a second). In practice, one can never measure a continuous variable to infinite precision, so continuous variables are sometimes approximated by discrete variables. A random variable X is also called *continuous* if its set of possible values is uncountable, and the chance that it takes any particular value is zero (in symbols, if $P(X = x) = 0$ for every real number x). A random variable is continuous if and only if its cumulative probability distribution function is a continuous function (a function with no jumps).

Contour plot A contour plot is the projection of a 3-dimensional surface onto a 2-dimensional plane.

Data contour.STA 5V°150c

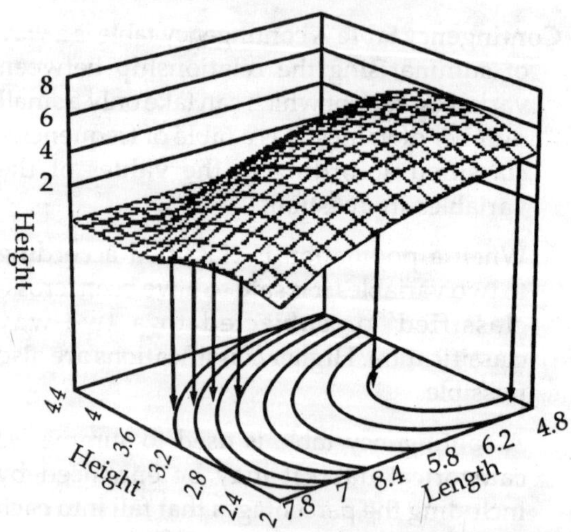

As compared to surface plots, they may be less effective to quickly visualize the overall shape of 3D data structures,

However, their main advantage is that they allow for precise examination and analysis of the shape of the surface (Contour Plots display a series of undistorted horizontal "cross sections" of the surface).

Contour/discrete raw data plot This sequential plot can be considered to be a 2D projection of the 3D Ribbons plot. Each data point in this plot is represented as a rectangular region, with different colors and/or patterns corresponding to the values (or range of values of the data points (the ranges are described in the legend). Values within each series are presented along the X-axis, with each series plotted along the Y-axis.

Contrapositive If *p* and *q* are two logical propositions, then the *contrapositive* of the proposition (*p* implies *q*) is the proposition ((not *q*) implies (not *p*)). The contrapositive is logically equivalent to the original proposition.

Control There are at least three senses of "control" in statistics: a member of the *control group*, to whom no treatment is given; a controlled experiment, and to control for a possible confounding variable.

Control for a variable To control for a variable is to try to separate its effect from the treatment effect, so it will not confound with the treatment. There are many methods that try to control for variables. Some are based on matching individuals between treatment and control; others use assumptions about the nature of the effects of the variables to try to model the effect mathematically, for example, using regression.

Control group The subjects in a controlled experiment who do not receive the treatment.

Control variable A variable that specifies which are the partial tables in a tabulation of three variables.

Controlled experiment An experiment that uses the method of comparison to evaluate the effect of a treatment by comparing treated subjects with a control group, who do not receive the treatment.

Controlled, randomized experiment A controlled experiment in which the assignment of subjects to the treatment group or control group is done at random, for example, by tossing a coin.

Convenience sample A sample drawn because of its convenience; not a probability sample. For example, I might take a sample of opinions in Berkeley (where I live) by just asking my 10 nearest neighbors. That would be a sample of convenience, and would be unlikely to be representative of all of Berkeley. Samples of convenience are not typically representative, and it is not typically possible to quantify how unrepresentative results based on samples of convenience will be.

Converge, convergence A sequence of numbers $x_1, x_2, x_3 \ldots$ *converges* if there is a number *x* such that for any number E>0, there is a number *k* (which can depend on E) such that $|x_j - x| < E$ whenever $j > k$. If such a number *x* exists, it is called the limit of the sequence $x_1, x_2, x_3 \ldots$.

Convergence in probability A sequence of random variables $X_1, X_2, X_3 \ldots$ *converges in probability* if there is a random variable X such that for any number E>0, the sequence of numbers

$P(|X_1 - X| < e), P(|X_2 - X| < e), P(|X_3 - X| < e),.$

converges to 100%.

Converse.

If *p* and *q* are two logical propositions, then the *converse* of the proposition (*p* implies *q*) is the proposition (*q* implies *p*).

Cook's distance This is another measure of impact of the respective case on the regression equation. It indicates the difference between the computed B values and the values one would have obtained, had the respective case been excluded. All distances should be of about equal magnitude; if not, then there is reason to believe that the respective case(s) biased the estimation of the regression coefficients.

See also, standard residual value, Mahalanobis distance, and deleted residual.

Correct Number of cases correctly classified.

Correlated Two (or more) quantities that change together in a consistent manner. Thus, if the value of one variable is known, the other can be immediately determined from their relationship.

Correlation Correlation is a measure of the relation between two or more variables. Correlation coefficients can range from -1.00 to +1.00. The value of -1.00 represents a perfect negative correlation while a value of +1.00 represents a perfect positive correlation. A value of 0.00 represents a lack of correlation.

Correlation coefficient A correlation coefficient is a number between -1 and 1 which measures the degree to which two variables are linearly related. If there is perfect linear relationship with positive slope between the two variables, we have a correlation coefficient of 1; if there is positive correlation, whenever one variable has a high (low) value, so does the other. If there is a perfect linear relationship with negative slope between the two variables, we have a correlation coefficient of -1; if there is negative correlation, whenever one variable has a high (low) value, the other has a low (high) value. A correlation coefficient of 0 means that there is no linear relationship between the variables.

There are a number of different correlation coefficients that might be appropriate depending on the kinds of variables being studied.

See also Pearson's Product Moment Correlation Coefficient.

See also Spearman Rank Correlation Coefficient.

Correlation value (association rules) When applying (in data or text mining) algorithms for deriving association rules of the general form If Body then Head (e.g., If (Car=Porsche and Age<20) then (Risk=High and Insurance=High)), a Correlation value can be computed as the support value for the rule, divided by the square root of the product of the support values for the Body and Head computed separately.

Correlational statistics Statistics that determine the strength and direction of the relationship between two variables.

Correspondence analysis Correspondence analysis is a descriptive/exploratory technique designed to analyze simple two-way and multi-way tables containing some measure of correspondence between the rows and columns. The results provide information which is similar in nature to those produced by Factor Analysis techniques, and they allow one to explore the structure of categorical variables included in the table. The most common kind of table of this type is the two-way frequency crosstabulation table.

Count The number of times cases with a particular combination of attributes occurs in a dataset. For example, the count of unemployed men within a dataset might equal 421.

Countable set A set is countable if its elements can be put in one-to-one correspondence with a subset of the integers. For example, the sets {0, 1, 7, -3}, {red, green, blue}, { . . . ,-2, -1, 0, 1, 2, . . . }, {straight, curly}, and the set of all fractions, are countable. If a set is not countable, it is uncountable. The set of all real numbers is *uncountable.*

Covariance A measure which indicates how the values of two continuos variables vary together. Obtained by dividing the sum of the cross-products by one less than the number of cases.

Cover A confidence interval is said to *cover* if the interval contains the true value of the parameter. Before the data are collected, the chance that the confidence interval will contain the parameter value is the coverage probability, which equals the confidence level after the data are collected and the confidence interval is actually computed.

Coverage probability The *coverage probability* of a procedure for making confidence intervals is the chance that the procedure produces an interval that covers the truth.

Cramer's V A measure of association appropriate for measuring the strength of a relationship between two variables, one or both of which has more than two categories.

Criterion The variable considered the dependent variable in a correlational study that is not a true experiment.

Critical region The critical region CR, or rejection region RR, is a set of values of the test statistic for which the null hypothesis is rejected in a hypothesis test; that is, the sample space for the test statistic is partitioned into two regions; one region (the critical region) will lead us to reject the null hypothesis H0', the other not. So, if the observed value of the test statistic is a member of the critical region, we conclude 'reject H0'; if it is not a member of the critical region then we conclude 'do not reject H0.

See also critical value. See also test statistic.

Critical value A number that the test statistic needs to exceed in order to reject the null hypothesis. In some situations, the null hypothesis is rejected for small values of the test statistic, and in these cases the test statistic needs to be less than the critical value. An alternative convention is to include the critical value itself in the rejection region. Since the test statistic is rarely exactly equal to the critical value, this alternative convention differs inconsequentially from the other.

Cross entropy Error functions based on information-theoretic measures, and particularly appropriate for classification networks. There are two versions, for single-output networks and multiple-output networks; these should be combined with the logistic and softmax activation functions respectively.

Cross entropy (single & multiple) This error is the sum of the products of the target value and the logarithm of the error value on each output unit. There are two versions: one for single-output (two-class) networks, the other for multiple-output networks. The cross-entropy error function is specially designed for classification problems, where it is used in combination with the logistic (single output) or softmax (multiple output) activation functions in the output layer of the network. This is equivalent to maximum likelihood estimation of the network weights. An MLP with no hidden layers, a single output unit, and cross entropy error function is equivalent to a standard logistic regression function (logit or probit classification).

Cross over design An experimental design involving two treatments in which every subject receives both treatments at different points in time, such that one group of subjects receives the treatments in one sequence, and another group in the opposite sequence. If both groups have the same number of subjects, the design is called balanced.

Cross product Obtained by multiplying the deviations from the means of pairs of values of two variables. The cross-products are obtained as a first step in the calculation of the covariance.

Cross sectional Adjective describing an observational study in which observations are taken at a particular point in time.

Cross sectional study A cross-sectional study compares different individuals to each other at the same time—it looks at a cross-section of a population. The differences between those individuals can confound with the effect being explored. For example, in trying to determine the effect of age on sexual promiscuity, a cross-sectional study would be likely to confound the effect of age with the effect of the mores the subjects were taught as children: the older individuals were probably raised with a very different attitude towards promiscuity than the younger subjects. Thus it would be imprudent to attribute differences in promiscuity to the aging process. *C.f.* longitudinal study.

Cross validation Cross-validation refers to the process of assessing the predictive accuracy of a model in a test sample (sometimes also called a cross-validation sample) relative to its predictive accuracy in the learning sample from which the model was developed. Ideally, with a large sample size, a proportion of the cases (perhaps one-half or two-thirds) can be designated as belonging to the learning sample and the remaining cases can be designated as belonging to the test sample. The model can be developed using the cases in the learning sample, and its predictive accuracy can be assessed using the cases in the test sample. If the model performs as well in the test sample as in the learning sample, it is said to cross-validate well, or simply to cross-validate.

A variety of techniques have been developed for performing cross-validation with small sample sizes by constructing test samples and learning samples which are partly but not wholly independent.

Cross verification The same as Cross-Validation. In the context of neural networks, the use of an auxiliary set of data (the verification set) during iterative training. While the training set is used to adjust the network weights, the verification set maintains an independent check that the neural network is learning to generalize.

Crosstabulations (tables, multi way tables) A crosstabulation table is a combination of two (or more) frequency tables arranged such that each cell in the resulting table represents a unique combination of specific values of crosstabulated variables. Thus, crosstabulation allows us to examine frequencies of observations that belong to specific combinations of categories on more than one variable. For example, the following simple ("two-way") table shows how many adults vs. children selected "cookie A" vs. "cookie B" in a taste preference test:

	COOKIE: A	COOKIE: B	
AGE: ADULT	50	0	50
AGE: CHILD	0	50	50
	50	50	100

By examining these frequencies, we can identify relations between crosstabulated variables (e.g., children clearly prefer "cookie B"). Only categorical (nominal) variables or variables with a relatively small number of different meaningful values should be crosstabulated. Note that in the cases where we do want to include a continuous variable in a crosstabulation (e.g., income), we can first recode it into a particular number of distinct ranges (e.g., low, medium, high).

C-SVM classification A support vector machine implementation for solving multi-classification prediction problems.

Cubic spline smoother The cubic spline scatterplot smoother is a special smoothing technique for 2D scatterplots, which generally produces a smooth generalization of the relationship between the two variables in the scatterplot. The cubic spline smoother is often used in generalized additive models, to estimate the unspecific (non-parametric) function of the predictor variables that best predicts the (transformed) dependent variable values. Computational details regarding the cubic spline smoother, and comparisons to other smoothing algorithms, can be found in Hastie and Tibshirani, 1990, and Schimek, 2000.

Cumulative distribution function All random variables (discrete and continuous) have a cumulative distribution function. It is a function giving the probability that the random variable X is less than or equal to x, for every value x.

Formally, the cumulative distribution function F(x) is defined to be:

$F(x) = P(X \leq x)$ for $-\infty < x < \infty$ c.d.f.

For a discrete random variable, the cumulative distribution function is found by summing up the probabilities as in the example below.

For a continuous random variable, the cumulative distribution function is the integral of its probability density function.

Example

Discrete case : Suppose a random variable X has the following probability distribution p(xi):

xi 0 1 2 3 4 5

P(xi) 1/32 5/32 10/32 10/32 5 / 32 1/32

This is actually a binomial distribution: Bi(5,1/2) or B(5,1/2). The cumulative distribution function F(x) is then:

xi 0 1 2 3 4 5

F(xi) 1/32 6/32 16/32 26/32 31 / 32 32/32.

F(x) does not change at intermediate values. For example: F (1.3) = F(1.86) = F(1) = 6/32.

Cumulative probability distribution function (CDF) The cumulative distribution function of a random variable is the chance that the random variable is less than or equal to *x*, as a function of *x*. In symbols, if *F* is the cdf of the random variable X, then *F*(*x*) = P(X <= *x*). The cumulative distribution function must tend to zero as *x* approaches minus infinity, and must tend to unity as *x* approaches infinity. It is a positive function, and increases monotonically: if *y* > *x*, then *F*(*y*) >= *F*(*x*). The cumulative distribution function completely characterizes the probability distribution of a random variable.

Curse of dimensionality The term curse of dimensionality generally refers to the difficulties involved in fitting models, estimating parameters, or optimizing a function in many dimensions, usually in the context of neural networks. As the dimensionality of the input data space (i.e., the number of predictors) increases, it becomes exponentially more difficult to find global optima for the parameter space, i.e., to fit models. In practice, the complexity of neural networks becomes unmanageable when the number of inputs into the neural network exceeds a few hundreds or even less, depending on the complexity of the respective neural network architecture.

Hence, it is simply a practical necessity to pre-screen and preselect from among a large set of input (predictor) variables those that are of likely utility for predicting the outputs (dependent variables) of interest.

See also, Feature Selection.

Cyclical component We want to increase our understanding of a time series by picking out its main features. One of these main features is the cyclical component. Descriptive techniques may be extended to forecast (predict) future values.

In weekly or monthly data, the cyclical component describes any regular fluctuations.

It is a non-seasonal component which varies in a recognisable cycle.

D

Daniell (or equal weight) window In Time Series, the Daniell window (Daniell 1946) is a weighted moving average transformation used to smooth the periodogram values. This transformation amounts to a simple (equal weight) moving average transformation of the periodogram values, that is, each spectral density estimate is computed as the mean of the m/2 preceding and subsequent periodogram values.

Data Numerical information about variables; the measurements or observations to be analyzed with statistical methods.

Data matrix The name given to the column by row organisation of numerical responses which results from coding survey questionnaires. The rows correspond to the cases and the columns contain the responses to each variable.

Data mean Average value of the target output variable.

Data mining StatSoft defines data mining as an analytic process designed to explore large amounts of (typically business or market related) data in search for consistent patterns and/or systematic relationships between variables, and then to validate the findings by applying the detected patterns to new subsets of data. Data mining uses many of the principles and techniques traditionally referred to as Exploratory Data Analysis (EDA).

Data preparation phase In Data Mining, the input data are often "noisy," containing many errors, and sometimes information in unstructured form (e.g., in Text Mining). For example, suppose you wanted to analyze a large database of information collected on-line via the web, based on voluntary responses of persons reviewing your web site (e.g., potential customers of a web-based retailer, who filled out suggestion forms). In those instances it is very important to first verify and "clean" the data in a data preparation phase, before applying any analytic procedures. For example, some individuals might enter clearly faulty information (e.g., age = 300), either by mistake or intentionally. If those types of data errors are not detected prior to the analysis phase of the data mining project, they can greatly bias the result, and potentially cause unjustified conclusions. Typically, during the data preparation phase, the data analyst applies "filters" to the data, to verify correct data ranges, and to delete impossible co-occurrences of values (e.g., Age=5; Retired=Yes).

Data reduction by decreasing the dimensionality (exploratory multivariate statistics) This interpretation of the term Data Reduction pertains to analytic methods (typically multivariate exploratory techniques such as Factor Analysis, Multidimensional Scaling, Cluster Analysis, Canonical Correlation, or Neural Networks) that involve reducing the dimensionality of a data set by extracting a number of underlying factors, dimensions, clusters, etc., that can account for the variability in the (multidimensional) data set. For example, in poorly designed questionnaires, all responses provided by the participants on a large number of variables (scales, questions, or dimensions) could be explained by a very limited number of "trivial" or artifactual factors. For example, two such underlying factors could be: 1. The respondent's attitude towards the study (positive or negative) and

2. The "social desirability" factor (a response bias representing a tendency to respond in a socially desirable manner).

Data reduction by unbiased decreasing of the sample size This type of Data Reduction is applied in exploratory graphical data analysis of extremely large data sets. The size of the data set can obscure an existing pattern (especially in large line graphs or scatterplots) due to the density of markers or lines. Then, it can be useful to plot only a representative subset of the data (so that the pattern is not hidden by the number of point markers) to reveal the otherwise obscured but still reliable pattern.

Data rotation (in 3d space) Changing the viewpoint for 3D scatterplots (e.g., simple, spectral, or space plots) may prove to be an effective exploratory technique since it can reveal patterns that are easily obscured unless you look at the "cloud" of data points from an appropriate angle.

Rotating or spinning a 3D graph will allow you to find the most informative location of the "viewpoint" for the graph.

Data S.D Standard deviation of the target output variable.

Data warehousing StatSoft defines data warehousing as a process of organizing the storage of large, multivariate data sets in a way that facilitates the retrieval of information for analytic purposes.

For more information, see Data Warehousing.

Decision rule A rule for comparing the outcome value of alpha with a nominal alpha criterion level (such as 0.05). An outcome value smaller (more extreme) than the nominal alpha criterion level leads to a decision of statistical significance of the finding that the test statistic has a value other than its (null-) hypothesised value.

Deck of cards A standard deck of playing cards contains 52 cards, 13 each of four suits: spades, hearts, diamonds, and clubs. The thirteen cards of each suit are {ace, 2, 3, 4, 5, 6, 7, 8, 9, 10, jack, queen, king}. The *face* cards are {jack, queen, king}. It is typically assumed that if a deck of cards is shuffled well, it is equally likely to be in each possible ordering. There are 52! (52 factorial) possible orderings.

Deduction A method of analysis that proceeds by formulating a theory and testing it with data.

Define The Define phase is concerned with the definition of project goals and boundaries, and the identification of issues that need to be addressed to achieve the higher sigma level.

Degrees of freedom Used in slightly different senses throughout the study of statistics, Degrees of Freedom were first introduced by Fisher based on the idea of degrees of freedom in a dynamical system (e.g., the number of independent co-ordinate values which are necessary to determine it). The degrees of freedom of a set of observations are the number of values which could be assigned arbitrarily within the specification of the system. For example, in a sample of size n grouped into k intervals, there are k-1 degrees of freedom, because k-1 frequencies are specified while the other one is specified by the total size n. Thus in a p by q contingency table with fixed marginal totals, there are (p-1)(q-1) degrees of freedom. In some circumstances the term degrees of freedom is used to denote the number of independent comparisons which can be made between the members of a sample.

Deleted residual The deleted residual is the residual value for the respective case, had it not been included in the regression analysis, that is, if one would exclude this case from all computations. If the deleted residual differs greatly from the respective standardized residual value, then this case is possibly an outlier because its exclusion changed the regression equation.

See also, standard residual value, Mahalanobis distance, and Cook's distance.

Delta bar delta Delta-bar-Delta is an alternative to back propagation, which is sometimes more efficient, although it can be more inclined to stick in local minima than back propagation. Unlike quick propagation, it tends to be quite stable.

Like quick propagation, Delta-bar-Delta is a batch algorithm: the average error gradient across all the training cases is calculated on each epoch, then the weights are updated once at the end of the epoch.

Delta-bar-Delta is inspired by the observation that the error surface may have a different gradient along each weight direction, and that consequently each weight should have its own learning rate (i.e. step size).

In Delta-bar-Delta, the individual learning rates for each weight are altered on each epoch to satisfy two important heuristics:

If the derivative has the same sign for several iterations, the learning rate is increased (the error surface has a low curvature, and so is likely to continue sloping the same way for some distance);

If the sign of the derivative alternates for several iterations, the learning rate is rapidly decreased (otherwise the algorithm may oscillate across points of high curvature).

To satisfy these heuristics, Delta-bar-Delta has an initial learning rate used for all weights on the first epoch, an increment factor added to learning rates when the derivative does not change sign, and a decay rate multiplied by the learning rates when the derivative does change sign. Using linear growth and exponential decay of learning rates contributes to stability.

The algorithm could still be prone to poor behavior on noisy error surfaces, where the derivative changes sign rapidly even within an overall downward trend. Consequently, the increase or decrease of learning rate is actually based on a smoothed version of the derivative.

Denominator synthesis A method developed by Satterthwaite (1946) which finds the linear combinations of sources of random variation that serve as appropriate error terms for testing the significance of the respective effect of interest in mixed-model ANOVA/ANCOVA designs.

Density function See probability density function.

Density scale The vertical axis of a histogram has units of percent per unit of the horizontal axis. This is called a density scale; it measures how "dense" the observations are in each bin. See also probability density.

Dependent events Events such that the outcome of one has an effect on the probability of the outcome of the other.

Dependent events, dependent random variables Two events or random variables are dependent if they are not independent.

Dependent variable A dependent variable is a variable whose values are predicted by another variable or variables, the independent variable(s). The dependent variable is usually the subject of primary interest in a study.

Deployment The concept of deployment in predictive data mining refers to the application of a model for prediction or classification to new data. After a satisfactory model of set of models have been identified (trained) for a particular application, one usually wants to deploy those models so that predictions or predicted classifications can quickly be obtained for new data. For example, a credit card company may want to deploy a trained model or set of models (e.g., neural networks, meta-learner) to quickly identify transactions which have a high probability of being fraudulent.

Derivative free function minimization algorithms Nonlinear Estimation offers several general function minimization algorithms that follow different search strategies which do not depend on the second-order derivatives. These strategies are sometimes very effective for minimizing loss functions with local minima.

Derived variable A derived variable is one that has been created out of the responses from the original variables in the datafile. Thus one may derive the variable agegroup from data collected about age.

Descriptive statistics Numerical data that describe phenomena.

Design matrix In general linear models and generalized linear models, the design matrix is the matrix X for the predictor variables which is used in solving the normal equations. X is a matrix, with 1 row for each case and 1 column for each coded predictor variable in the design, whose values identify the levels for each case on each coded predictor.

See also general linear model, generalized linear model.

Desirability profiles The relationship between predicted responses on one or more dependent variables and the desirability of responses is called the desirability function. Profiling the desirability of responses involves, first, specifying the desirability function for each dependent variable, by assigning predicted values a score ranging from 0 (very undesirable) to 1 (very desirable). The individual desirability scores for the predicted values for each dependent variable are then combined by computing their geometric mean. Desirability profiles consist of a series of graphs, one for each independent variable, of overall desirability scores at different levels of one independent variable, holding the levels of the other independent variables constant at specified values. Inspecting the desirability profiles can show which levels of the predictor variables produce the most desirable predicted responses on the dependent variables.

Detrended normal probability plot In this plot each value (x_j) is standardized by subtracting the mean and dividing by the respective standard deviation (s). The detrended normal probability value z_j for the jth value (rank) in a variable with N observations is computed as:

$$z_j = \Phi^{-1} [(3*j-1)/(3*N+1)] - (x_j\text{-mean})/s$$

Where Φ^{-1} is again the inverse normal cumulative distribution function.

Detrended probability plots This type of graph is used to evaluate the normality of the distribution of a variable, that is, whether and to what extent the distribution of the variable follows the normal distribution. The selected variable will be plotted in a scatterplot against the values "expected from the normal distribution." This plot is constructed in the same way as the standard normal probability plot, except that before the plot is generated, the linear trend is removed. This often "spreads out" the plot, thereby allowing the user to detect patterns of deviations more easily.

Deviance To evaluate the goodness of fit of a generalized linear model, a common statistic that is computed is the so-called Deviance statistic. It is defined as:

Deviance = -2 * (Lm – Ls)

where Lm denotes the maximized log-likelihood value for the model of interest, and Ls is the log-likelihood for the saturated model, i.e., the most complex model given the current distribution and link function .

Deviance residuals After fitting a generalized linear model to the data, to check the adequacy of the respective model, one usually computes various residual statistics. The deviance residual is computed as:

$r_D = \text{sign}(y-\mu)\text{sqrt}(d_i)$

Where $?d_i = D$, and D is the overall deviance measure of discrepancy of a generalized linear model. Thus, the deviance statistic for an observation reflects it's contribution to the overall goodness of fit (deviance) of the model.

Deviation A deviation is the difference between a datum and some reference value, typically the mean of the data. In computing the SD, one finds the rms of the deviations from the mean, the differences between the individual data and the mean of the data.

Deviation assignment algorithms These algorithms assign deviations to the radial units in certain network types. The deviation is multiplied by the distance between the unit's exemplar vector and the input vector, to determine the unit's output. In essence, the deviation gives the size of the cluster represented by a radial unit.

Deviation assignment algorithms are used after radial centers have been set; see Radial Sampling and K Means.

Deviation plots 3d Data (representing the X, Y, and Z coordinates of each point) in this type of graph are represented in 3D space as "deviations" from a specified base-level of the Z-axis.

Deviation plots are similar to space plots. As compared to space plots, however in deviation plots the "deviations plane" is "invisible" and not marked by the location of the X-Y axes (those axes are always fixed in the standard bottom position). Deviation plots may help explore the nature of 3D data sets by displaying them in the form of deviations from arbitrary (horizontal) levels. Such "cutting" methods can help identify interactive relations between variables.

DFFITS Several measures have been given for testing for leverage and influence of a specific case in regression (including studentized residuals, studentized deleted residuals, DFFITS, and standardized DFFITS). Belsley et al. (1980) have suggested DFFITS, a measure which gives greater weight to outlying observations than Cook's distance. The formula for DFFITS is

$DFFIT_i = \tilde{h}_i e_i/(1-\tilde{h}_i)$

where e_i is the error for the ith case

h_i is the leverage for the ith case

and $\tilde{h}_i = 1/N + h_i$.

Dichotomous A variable with only two categories, such as male and female, is called dichotomous.

Diehard suite of tests and random number generation Many areas of statistical analysis, research, and simulation rely on the quality of random number generators. Most programs for statistical data analysis contain a function for generating uniform random numbers. A recent review of statistical packages (McCullough, 1998, 1999) that appeared in The American Statistician tested the random number generators of several programs using the so-called diehard suite of tests (Marsaglia, 1998). Diehard applies various methods of assembling and combining uniform random numbers, and then performs statistical tests that are expected to be nonsignificant; this suite of tests has become a standard method of evaluating the quality of uniform random number generator routines.

Difference of means A test statistic of intuitive appeal for measuring difference in location between two samples with interval-scale data. Employing this test statistic in an exact test defines the pitman permutation tests(1 or 2).

Differencing Differencing is a popular and effective method of removing trend from a time series. This provides a clearer view of the true underlying behaviour of the series.

Differencing (in time series) In this Time Series transformation, the series will be transformed as: X=X-X(lag). After differencing, the resulting series will be of length N-lag (where N is the length of the original series).

Dimensionality reduction Data Reduction by decreasing the dimensionality (exploratory multivariate statistics). This interpretation of the term Data Reduction pertains to analytic methods (typically multivariate exploratory techniques such as Factor Analysis, Multidimensional Scaling, Cluster Analysis, Canonical Correlation, or Neural Networks) that involve reducing the dimensionality of a data set by extracting a number of underlying factors, dimensions, clusters, etc., that can account for the variability in the (multidimensional) data set.

Directional test A test of the prediction that one value is higher than another; also called a one-tailed test.

Discordant If there are two cases, A and B, and case A has a higher value on one variable than case B, and case A has a lower value on another variable than case B, the two cases are discordant for the two variables. They are also discordant if case A has a lower value on one variable while case B has a higher value on the other.

Discrepancy function A numerical value that expresses how badly a structural model reproduces the observed data. The larger the value of the discrepancy function, the worse (in some sense) the fit of model to data. In general, the parameter estimates for a given model are selected to make a discrepancy function as small as possible.

The discrepancy functions employed in structural modeling all satisfy the following basic requirements:

They are non-negative, i.e., always greater than or equal to zero.

They are zero only if fit is perfect, i.e., if the model and parameter estimates perfectly reproduce the observed data.

The discrepancy function is a continuous function of the elements of S, the sample covariance matrix, and $\Sigma(\theta)$, the "reproduced" estimate of S obtained by using the parameter estimates and structural model.

Discrete data A set of data is said to be discrete if the values / observations belonging to it are distinct and separate, i.e. they can be counted (1,2,3,....). Examples might include the number of kittens in a litter; the number of patients in a doctors surgery; the number of flaws in one metre of cloth; gender (male, female); blood group (O, A, B, AB).

Compare continuous data.

Discrete distribution A probability distribution of some statistic, based upon an algebraic formula or upon re-randomisation or upon actual data, in which the cumulative probability increases in non-infinitesmal steps corresponding to non-infinitesmal weight associated with possible values of the statistic in question. This situation is characteristic of randomisation distributions, and also of test statistics which are essentially discrete.

Discrete random variable A discrete random variable is one which may take on only a countable number of distinct values such as 0,1,2,3,4,........ Discrete random variables are usually (but not necessarily) counts. If a random variable can take only a finite number of distinct values, then it must be discrete. Examples of discrete random variables include the number of children in a family, the Friday night attendance at a cinema, the number of patients in a doctor's surgery, the number of defective light bulbs in a box of ten.

Compare continuous random variable.

Discrete variable A quantitative variable whose set of possible values is countable. Typical examples of discrete variables are variables whose possible values are a subset of the integers, such as the number of vehicles that cross a particular bridge on a given day, the number of people in a family, ages rounded to the nearest year, *etc.* Discrete variables are "chunky." *C.f.* continuous variable. A discrete random variable is one whose set of possible values is countable. A random variable is discrete if and only if its cumulative probability distribution function is a stair-step function; *i.e.*, if it is piecewise constant and only increases by jumps.

Discriminant function analysis Discriminant function analysis is used to determine which variables discriminate between two or more naturally occurring groups (it is used as either a hypothesis testing or exploratory method). For example, an educational researcher may want to investigate which variables discriminate between high school graduates who decide 1. To go to college,

2. To attend a trade or professional school, or

3. To seek no further training or education. For that purpose the researcher could collect data on numerous variables prior to students' graduation. After graduation, most students will naturally fall into one of the three categories. Discriminant Analysis could then be used to determine which variable(s) are the best predictors of students' subsequent educational choice (e.g., IQ, GPA, SAT).

Disjoint occurrence Both outcomes unable to happen at the same time.

Disjoint or mutually exclusive events Two events are disjoint or mutually exclusive if the occurrence of one is incompatible with the occurrence of the other; that is, if they can't both happen at once (if they have no outcome in common). Equivalently, two events are disjoint if their intersection is the empty set.

Disjoint or mutually exclusive sets Two sets are disjoint or mutually exclusive if they have no element in common. Equivalently, two sets are disjoint if their intersection is the empty set.

Dispersion The data values in a sample are not all the same. This variation between values is called dispersion.

When the dispersion is large, the values are widely scattered; when it is small they are tightly clustered. The width of diagrams such as dot plots, box plots, stem and leaf plots is greater for samples with more dispersion and vice versa.

There are several measures of dispersion, the most common being the standard deviation. These measures indicate to what degree the individual observations of a data set are dispersed or 'spread out' around their mean.

In manufacturing or measurement, high precision is associated with low dispersion.

Distribution The distribution of a set of numerical data is how their values are distributed over the real numbers. It is completely characterized by the empirical distribution function. Similarly, the probability distribution of a random variable is completely characterized by its probability distribution function. Sometimes the word "distribution" is used as a synonym for the empirical distribution function or the probability distribution function.

Distribution function, empirical The empirical (cumulative) distribution function of a set of numerical data is, for each real value of x, the fraction of observations that are less than or equal to x. A plot of the empirical distribution function is an uneven set of stairs. The width of the stairs is the spacing between adjacent data; the height of the stairs depends on how many data have exactly the same value. The distribution function is zero for small enough (negative) values of x, and is unity for large enough values of x. It increases monotonically: if $y > x$, the empirical distribution function evaluated at y is at least as large as the empirical distribution function evaluated at x.

Document frequency The document frequency is a useful statistic computed for individual words or terms in the context of text mining. It denotes the number of documents in a collection of documents in which the respective word or term was found.

Dot plot A dot plot is a way of summarising data, often used in exploratory data analysis to illustrate the major features of the distribution of the data in a convenient form.

For nominal or ordinal data, a dot plot is similar to a bar chart, with the bars replaced by a series of dots. Each dot represents a fixed number of individuals. For continuous data, the dot plot is similar to a histogram, with the rectangles replaced by dots.

A dot plot can also help detect any unusual observations (outliers), or any gaps in the data set.

Double blind experiment In a double-blind experiment, neither the subjects nor the people evaluating the subjects knows who is in the treatment group and who is in the control group. This mitigates the placebo effect and guards against conscious and unconscious prejudice for or against the treatment on the part of the evaluators.

Double counting A mistake encountered in calculating the probability of at least one of several events occurring, when the events are not mutually exclusive. In this case, the addition rule does not apply.

DoubleY histograms The Double-Y histogram can be considered to be a combination of two separately scaled multiple histograms. Two different series of variables can be selected. A frequency distribution for each of the selected variables will be plotted but the frequencies of the variables entered into the first list (called Left-Y variables) will be plotted against the left-Y axis, whereas the frequencies of the variables entered into the second list (called Right-Y variables) will be plotted against the right-Y axis. The names of all variables from the two lists will be included in the legend followed by a letter L or R, denoting the Left-Y and Right-Y axis, respectively.

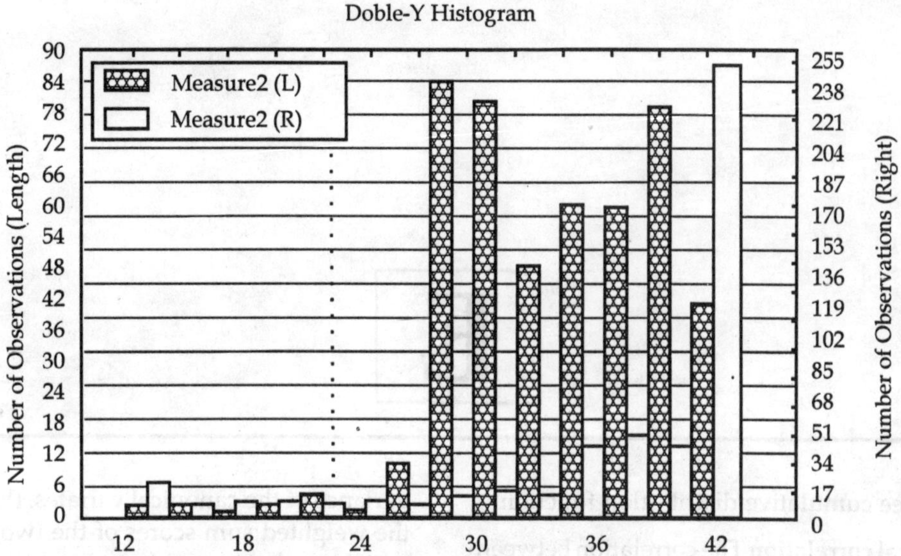

This graph is useful to compare distributions of variables with different frequencies.

Drill down analysis The concept of drill-down analysis applies to the area of data mining, to denote the interactive exploration of data, in particular of large databases. The process of drill-down analyses begins by considering some simple break-downs of the data by a few variables of interest (e.g., Gender, geographic region, etc.). Various statistics, tables, histograms, and other graphical summaries can be computed for each group. Next one may want to "drill-down" to expose and further analyze the data "underneath" one of the categorizations, for example, one might want to further review the data for males from the mid-west. Again, various statistical and graphical summaries can be computed for those cases only, which might suggest further break-downs by other variables (e.g., income, age, etc.). At the lowest ("bottom") level are the raw data: For example, you may want to review the addresses of male customers from one region, for a certain income group, etc., and to offer to those customers some particular services of particular utility to that group.

Dummy variable (in regression) In regression analysis we sometimes need to modify the form of non-numeric variables, for example sex, or marital status, to allow their effects to be included in the regression model. This can be done through the creation of dummy variables whose role it is to identify each level of the original variables separately.

Duncan's test This post hoc test (or multiple comparison test) can be used to determine the significant differences between group means in an analysis of variance setting. Duncan's test, like the Newman-Keuls test, is based on the range

Dunnett's test This post hoc test (or multiple comparison test) can be used to determine the significant differences between a single control group mean and the remaining treatment group means in an analysis of variance setting. Dunnett's test is considered to be one of the least conservative post hoc tests.

DV DV stands for Dependent Variable. See also dependent vs. Independent variables.

E

ECDF See cumulative distribution function

Ecological correlation The correlation between averages of groups of individuals, instead of individuals. Ecological correlation can be misleading about the association of individuals.

Effective hypothesis decomposition When in a factorial ANOVA design there are missing cells, then there is ambiguity regarding the specific comparisons between the (population, or least-squares) cell means that constitute the main effects and interactions of interest.

In addition, for sigma restricted models (e.g., in General Regression Models; some software offers the user a choice between the sigma restricted and overparameterized models), we propose a Type VI sums of squares option; this approach is identical to what is described as the effective hypothesis method in Hocking (1996).

Efficient score statistic See score statistic.

Eigen values Eigen values are computed when extracting the canonical roots. These can be interpreted as the proportion of variance accounted for by the correlation between the respective canonical variates. Note that the proportion here is computed relative to the variance of the canonical variates, that is, of the weighted sum scores of the two sets of variables; the eigenvalues do not tell how much variability is explained in either set of variables. As many eigenvalues are computed as there are canonical roots, that is, as many as the minimum number of variables in either of the two sets.

Elaboration A method for exploring and testing ideas about causal relationships between three or more variables It involves examining tables of two variables, controlling for a third.

Element of a set See member.

Elementary event Also called simple event, atomic event, or outcome.

An event that, in a specific context, will not be refined by specifying additional information.

Ellipse, prediction interval (area) and range. prediction interval (area) ellipse This type of ellipse is useful for establishing confidence intervals for the prediction of single new observations (prediction intervals). Such bivariate confidence or control limits are, for example, often used in the context of multivariate control charts for industrial quality control.

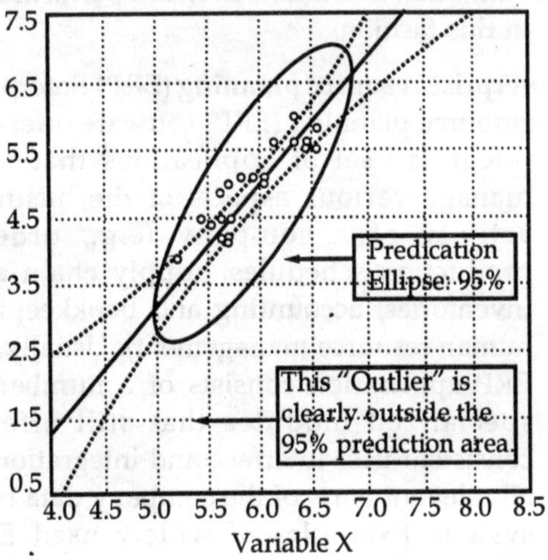

The ellipse is determined based on the assumption that the two variables follow the bivariate normal distribution. The orientation of this ellipse is determined by the sign of the linear correlation between two variables (the longer axis of the ellipse is superimposed on the regression line). The probability that the values will fall within the area marked by the ellipse is determined by the value of the coefficient that defines the ellipse (e.g., 95%).

EM clustering The EM clustering algorithm is a clustering technique similar in purpose to the k-means clustering method. The EM algorithm performs clustering by fitting a mixture of distributions to the data.

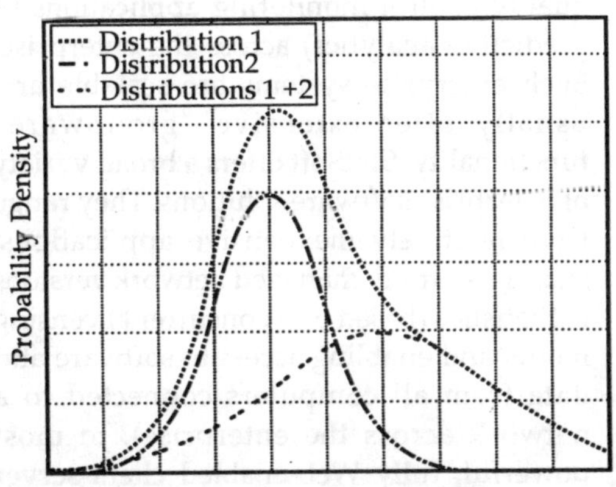

Only the mixture (sum) of the two normal distributions, with different means and standard deviations, would be observed (e.g., suppose clustering was performed based on a single continuous variable). The goal of EM clustering is to estimate the means and standard deviations for each cluster, so as to maximize the likelihood of the observed data (distribution).

Empirical law of averages The Empirical Law of Averages lies at the base of the frequency theory of probability. This law, which is, in fact, an assumption about how the world works, rather than a mathematical or physical law, states that if one repeats a random experiment over and over, independently and under "identical" conditions, the fraction of trials that result in a given outcome converges to a limit as the number of trials grows without bound.

Empirical rule A rule that is founded on observation, without a theoretical basis. Or a "rule of thumb."

Empty set The empty set, denoted {} or Ø, is the set that has no members.

Endogenous variable An endogenous variable is a variable that appears as a dependent variable in at least one equation in a structural model. In a path diagram, endogenous variables can be recognized by the fact that they have at least one arrow pointing to them.

Endpoint convention In plotting a histogram, one must decide whether to include a datum that lies at a class boundary with the class interval to the left or the right of the boundary. The rule for making this assignment is called an *endpoint convention.* The two standard endpoint conventions are 1. To include the left endpoint of all class intervals and exclude the right, except for the rightmost class interval, which includes both of its endpoints, and

2. To include the right endpoint of all class intervals and exclude the left, except for the leftmost interval, which includes both of its endpoints.

Ensemble architecture This is of the form I:[N]:O, where I is the number of input variable, O the number of output variables, and N the number of members of the ensemble.

Ensembles Ensembles are collections of neural networks that cooperate in performing a prediction. There are a number of uses for ensembles:

Ensembles can conveniently group together networks that provide predictions for related variables without requiring that all those variables be combined into a single network. Multiple output networks often suffer from cross-talk in the hidden neurons, and make ineffective predictions. Using an ensemble, each output can be predicted separately.

Ensembles provide an important method to combat over-learning and improve generalization. Averaging predictions across models with different structures, and/or trained on different data subsets, can reduce model variance without increasing model bias. This is a relatively simple way to improve generalization. Ensembles therefore are particularly effective when combined with resampling. An important piece of theory shows that the expected performance of an ensemble is greater than or equal to the average performance of the members.

Ensembles report the average performance and error measures of their member networks. You can perform resampling experiments, and save the results to an ensemble. Then, these average measures give an unbiased estimate of an individual network's performance, if trained in the same fashion. It is standard practice to use resampling techniques such as cross validation to estimate network performance in this fashion.

Enterprise resource planning (ERP) Enterprise resource planning (ERP) software offers an integrated set of applications that help manage various aspects of the business activity of a company (e.g., orders, production schedules, supply chain and inventories, accounting and bookkeeping, human resource management). Usually, an ERP application consists of a number of specialized modules that still offer a consistent user interface and integration of all relevant parts of the company data base system. Examples of widely used ERP applications are products of SAP and Oracle.

Enterprise SPC Enterprise SPC is a groupware based process control system, designed to work in enterprise-wide environment and allowing engineers and supervisors to share data, chart specifications (and other QC criteria), reports, and database queries. Enterprise SPC systems always include central QC data bases and if properly integrated, they allow the managers to maintain quality standards for all products/ processes in a given corporation.

Enterprise wide software systems Enterprise-Wide Software Systems enable large groups of employees to share and disseminate information (e.g., access to databases) and analytic/computational procedures (e.g., quality control monitoring applications or predictive analytics) across the enterprise. Such enterprise systems are scalable and usually offer extensive "groupware" functionality. StatSoft offers a broad variety of enterprise software solutions. They range from relatively inexpensive applications, such as centrally managed network versions of Statistica (based on a concurrent licensing model and enabling access to software and data from all computers connected to a network across the enterprise), to most powerful, fully Web-enabled client-server systems, built on a distributed processing architecture. StatSoft enterprise Web-based solutions (designed for various industries and applications) are seamlessly integrated with multiple database systems where they can process data "in-place" (including StatSoft integrated data warehouse solutions, document management systems, and other options), and offer tools to custom design interactive portals for different categories of users to facilitate their access to specific information and functionality that each of these categories of users need.

Epistemic An interpretation of probability concerned with the degree of knowledge or belief one has about a fact. cf. aleatory

Epoch During iterative training of a neural network, an Epoch is a single pass through the entire training set, followed by testing of the verification set.

EPS In fitted function equations, EPS stands for epsilon (e), The random error associated with measurement.

EPSEM samples EPSEM samples are probability samples where each observation in the population has the same known probability of being selected into the sample (EPSEM stands for equal probability of selection method sampling). Epsem samples have certain desirable properties; for example, the simple formulas for computing means, standard deviations, and so on can be applied to estimate the respective parameters in the population.

Equivalent test statistic Within a randomisation set, it is possible that two different statistics may be inter-related in a manner which is provably monotonic irrespective of the data. In such a situation a randomisation test performed on either of these test statistics will necessarily have the same outcome in terms of alpha. If one of the statistics is of good descriptive validity whereas the other is simpler to compute, then a randomisation test upon the simpler statistic may be used in place of a test upon the descriptively more valid one, with corresponding savings in amount of computation required. An example of such equivalent test statistics occurs for the situation of comparison of levels of a single interval-scale variable between two groups. In this situation, the descriptively valid statistic, as defined for the pitman permutation test(1), is the difference of means, but simpler equivalent test statistics include the mean for one designated group, or (most simply) the total of scores in one designated group.

Error bars (2d box plots) In this style of 2D Box plots, the ranges or error bars are calculated from the data. The central tendency (e.g., median or mean), and range or variation statistics (e.g., min-max values, quartiles, standard errors, or standard deviations) are computed for each variable and the selected values are presented as error bars.

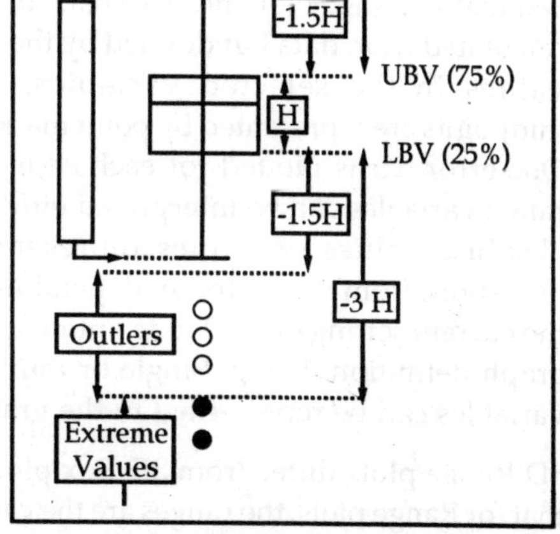

Error bars (2d range plots) In this style of 2D Range Plot, the ranges or error bars are defined by the raw values in the selected variables. The midpoints are represented by point markers. One range or error bar is plotted for each case. In the simplest instance, three variables need to be selected, one representing the mid-points, one representing the upper limits and one representing the lower limits.

Error bars (3d box plots) In this style of 3D Sequential Box Plot, the ranges of values of selected variables are plotted separately for groups of cases defined by values of a categorical (grouping) variable. The central tendency (e.g., median or mean), and range or variation statistics (e.g., min-max values, quartiles, standard errors, or standard deviations) are computed for each variable and for each group of cases and the selected values are presented as error bars.

3D Range plots differ from 3D Box plots in that for Range plots, the ranges are the values of the selected variables (e.g., one variable contains the minimum range values and another variable contains the maximum range values) while for Box plots the ranges are calculated from variable values (e.g., standard deviations, standard errors, or min-max value).

Error bars (3d range plots) In this style of 3D Sequential Range Plot, the error bars are not calculated from data but defined by the raw values in the selected variables. The midpoints are represented by point markers. One error bar is plotted for each case. The range variables can be interpreted either as absolute values or values representing deviations from the midpoint depending on the current setting of the Mode option in the graph definition dialog. Single or multiple variables can be represented in the graph.

3D Range plots differ from 3D Box plots in that for Range plots, the ranges are the values of the selected variables (e.g., one variable contains the minimum range values and another variable contains the maximum range values) while for Box plots, the ranges are calculated from variable values (e.g., standard deviations, standard errors, or min-max values).

Error function The error function is used in training the network and in reporting the error. The error function used can have a profound effect on the performance of training algorithms (Bishop, 1995).

Error mean Average error (residual between target and actual output values) of the output variable.

Error S.D Standard deviation of errors for the output variable.

Error sum of squares See residual sum of squares.

Estimable functions In general linear models and generalized linear models, if the X'X matrix (where X is the design matrix) is less than full rank, the regression coefficients depend on the particular generalized inverse used for solving the normal equations, and the regression coefficients will not be unique. When the regression coefficients are not unique, linear functions (f) of the regression coefficients having the form

$$f = Lb$$

where L is a vector of coefficients, will also in general not be unique. However, Lb for an L which satisfies

$$L = L(X'X)'X'X$$

is invariant for all possible generalized inverses, and is therefore called an estimable function.

See also general linear model, generalized linear model, design matrix, matrix rank, generalized inverse.

Estimate An estimate is an indication of the value of an unknown quantity based on observed data.

More formally, an estimate is the particular value of an estimator that is obtained from a particular sample of data and used to indicate the value of a parameter.

Example

Suppose the manager of a shop wanted to know the mean expenditure of customers in her shop in the last year. She could calculate the average expenditure of the hundreds (or perhaps thousands) of customers who bought goods in her shop, that is, the population mean. Instead she could use an estimate of this population mean by calculating the mean of a representative sample of customers. If this value was found to be £25, then £25 would be her estimate.

Estimate of sigma When the data set consists of multiple samples, such as data collected for the quality control chart, then one can compute two different indices of variability in the data. One is the regular standard deviation for all observations, ignoring the fact that the data consist of multiple samples; the other is to estimate the process's inherent variation from the within-sample variability. When the total process variability is used in the standard capability computations, the resulting indices are usually referred to as process performance indices (as they describe the actual performance of the process; common indices are P_p, P_r, and P_{pk}), while indices computed from the inherent variation (within-sample sigma) are referred to as capability indices (since they describe the inherent capability of the process; common indices are C_p, C_r, and C_{pk}).

See Process Capability Indices and Process Capability Analysis.

Estimation Estimation is the process by which sample data are used to indicate the value of an unknown quantity in a population.

Results of estimation can be expressed as a single value, known as a point estimate; or a range of values, known as a confidence interval.

Estimator An estimator is any quantity calculated from the sample data which is used to give information about an unknown quantity in the population. For example, the sample mean is an estimator of the population mean.

Estimators of population parameters are sometimes distinguished from the true value by using the symbol 'hat'. For example,

σ = true population standard deviation

$\hat{\sigma}$ = estimated (from a sample) population standard deviation

Example

The usual estimator of the population mean is $\hat{\mu} = \bar{X} = \frac{\sum X_i}{n}$

where n is the size of the sample and X1,X2,X3,.......,Xn are the values of the sample.

If the value of the estimator in a particular sample is found to be 5, then 5 is the estimate of the population mean μ.

Euclidean distance One can think of the independent variables (in a regression equation) as defining a multidimensional space in which each observation can be plotted. The Euclidean distance is the geometric distance in that multidimensional space. Note that Euclidean (and squared Euclidean) distances are computed from raw data, and not from standardized data.

Euler's E The base of the natural logarithm (numerical value: 2.71828182834905...), named after the Swiss mathematician Leonhard Euler (1707-1783).

Event An event is any collection of outcomes of an experiment.

Formally, any subset of the sample space is an event.

Any event which consists of a single outcome in the sample space is called an elementary or simple event. Events which consist of more than one outcome are called compound events.

Set theory is used to represent relationships among events. In general, if A and B are two events in the sample space S, then:

A∪B (A union B) = 'either A or B occurs or both occur'.

A∩B (A intersection B) = 'both A and B occur'.

A⊆B (A is a subset of B) = 'if A occurs, so does B'.

A' or $\bar{A}$ = 'event A does not occur'.

ϕ (the empty set) = an impossible event.

S (the sample space) = an event that is certain to occur.

Example

Experiment Rolling a dice once.

Sample space S = {1,2,3,4,5,6}.

Events A = 'score < 4' = {1,2,3}.

B = 'score is even' = {2,4,6}.

C = 'score is 7' = ϕ.

A∪B= 'the score is < 4 or even or both' = {1,2,3,4,6}.

A∩B= 'the score is < 4 and even' = {2}.

A' or $\bar{A}$ = 'event A does not occur' = {4,5,6}.

Exact binomial test A statistical test referring to the binomial distribution in its exact algebraic form, rather than through continuous approximations which are used especially where sample sizes are substantial.

Exact test(1) The characteristic of a re-randomisation test based upon exhaustive re-randomisation, that the value of alpha will be fixed irrespective of any random sampling of randomisations or upon any distributional assumptions. Notable examples are the exact binomial test, fisher test(1), the pitman permutation tests(1 and 2), and various non-parametric tests based upon ranked data.

Exact test(2) A test which yields an alpha value which does not depend upon the nominal alpha criterion value which may have been set for alpha. This is in contrast to the possible practice of producing only a yes/no decision with regard to a nominal alpha criterion value. Note that this reference to exactness is not (sic) the concern of the exact-stats initiative.

Exhaustive A collection of events $\{A_1, A_2, A_3, \ldots\}$ is *exhaustive* if at least one of them must occur; that is, if

$S = A_1 \cup A_2 \cup A_3 \cup$.

where S is the outcome space.

A collection of subsets *exhausts* another set if that set is contained in the] union of the collection.

Exhaustive re randomisation A series of samples from a randomisation set which is known to generate every randomisation. In particular, sampling which generates every randomisation exactly once.

Exogenous variable An exogenous variable is a variable that never appears as a dependent variable in any equation in a structural model. In a path diagram, exogenous variables can be recognized by the fact that they have no arrows pointing to them.

Expectation Also called expected value.

The mean value of the distribution of a random variable.

Expectation, expected value The expected value of a random variable is the long-term limiting average of its values in independent repeated experiments. The expected value of the random variable X is denoted EX or E(X). For a discrete random variable (one that has a countable number of possible values) the expected value is the weighted average of its possible values, where the weight assigned to each possible value is the chance that the random variable takes that value. One can think of the expected value of a random variable as the point at which its probability histogram would balance, if it were cut out of a uniform material. Taking the expected value is a linear operation: if X and Y are two random variables, the expected value of their sum is the sum of their expected values (E(X+Y) = E(X) + E(Y)), and the expected value of a constant *a* times a random variable X is the constant times the expected value of X (E(*a*×X) = *a*× E(X)).

Expected For a table of two variables, the count which would be obtained if there were no association between the two variables.

Expected counts The counts that would be obtained if a null hypothesis were true (usually the hypothesis is that there is no association between two variables).

Expected frequencies In contingency table problems, the expected frequencies are the frequencies that you would predict ('expect') in each cell of the table, if you knew only the row and column totals, and if you assumed that the variables under comparison were independent.

See also contingency table.

Expected residual life or average remaining lifetime The expected length of time until death for an individual who has survived to a specific age.

Expected value The expected value (or population mean) of a random variable indicates its average or central value. It is a useful summary value (a number) of the variable's distribution.

Stating the expected value gives a general impression of the behaviour of some random variable without giving full details of its probability distribution (if it is discrete) or its probability density function (if it is continuous).

Two random variables with the same expected value can have very different distributions. There are other useful descriptive measures which affect the shape of the distribution, for example variance.

The expected value of a random variable X is symbolised by E(X) or u.

If X is a discrete random variable with possible values x1, x2 ,x3 ,......., xn, and p(xi) denotes P(X = xi), then the expected value of X is defined by:

$$\mu = E(X) = \sum xi\, p(xi)$$

where the elements are summed over all values of the random variable X.

If X is a continuous random variable with probability density function f(x), then the expected value of X is defined by:

$$\mu = E(X) = \int x f(x) dx$$

Example

Discrete case : When a die is thrown, each of the possible faces 1,2,3,4,5,6 - the xi's - has a probability of 1/6 - the p(xi)'s - of showing. The expected value of the face showing is therefore:

$$\mu = E(X) = \sum xi\, p(xi)$$

= (1 x 1/6) +(2 x 1/6) +(3 x 1/6) +(4 x 1/6) +(5 x 1/6) +(6 x 1/6) = 3.5

Notice that, in this case, E(X) is 3.5, which is not a possible value of X.

Experiment A study in which the investigator performs some intervention. An experiment is any process or study which results in the collection of data, the outcome of which is unknown. In statistics, the term is usually restricted to situations in which the researcher has control over some of the conditions under which the experiment takes place. What distinguishes an experiment from an observational study is that in an experiment, the experimenter decides who receives the treatment.

Experimental design We are concerned with the analysis of data generated from an experiment. It is wise to take time and effort to organise the experiment properly to ensure that the right type of data, and enough of it, is available to answer the questions of interest as clearly and efficiently as possible. This process is called experimental design.

The specific questions that the experiment is intended to answer must be clearly identified before carrying out the experiment. We should also attempt to identify known or expected sources of variability in the experimental units since one of the main aims of a designed experiment is to reduce the effect of these sources of variability on the answers to questions of interest. That is, we design the experiment in order to improve the precision of our answers.

See Completely Randomised Design.

See Randomised Complete Block Design.

See Factorial Design.

Experimental design (DOE, industrial experimental design) In industrial settings, Experimental design (DOE) techniques apply analysis of variance principles to product development. The primary goal is usually to extract the maximum amount of unbiased information regarding the factors affecting a production process from as few (costly) observations as possible. In industrial settings, complex interactions among many factors that influence a product are often regarded as a "nuisance" (they are often of no interest; they only complicate the process of identifying important factors, and in experiments with many factors it would not be possible or practical to identify them anyway). Hence, if you review standard texts on experimentation in industry (Box, Hunter, and Hunter, 1978; Box and Draper, 1987; Mason, Gunst, and Hess, 1989; Taguchi, 1987) you will find that they will primarily discuss designs with many factors (e.g., 16 or 32) in which interaction effects cannot be evaluated, and the primary focus of the discussion is how to derive unbiased main effect (and, perhaps, two-way interaction) estimates with a minimum number of observations.

Experimental (or sampling) unit A unit is a person, animal, plant or thing which is actually studied by a researcher; the basic objects upon which the study or experiment is carried out. For example, a person; a monkey; a sample of soil; a pot of seedlings; a postcode area; a doctor's practice.

Explained variance The proportion of the variability in the data which is accounted for by the model (e.g., in Multiple Regression, ANOVA, Nonlinear Estimation, Neural Networks).

Explanation An independent variable explains a dependent variable if knowledge of the value of the independent variable provides good predictions of the value of the dependent variable for all cases.

Explanatory variable In regression, the explanatory or independent variable is the one that is supposed to "explain" the other. For example, in examining crop yield versus quantity of fertilizer applied, the quantity of fertilizer would be the explanatory or independent variable, and the crop yield would be the dependent variable. In experiments, the explanatory variable is the one that is manipulated; the one that is observed is the dependent variable.

Explicit deviation assignment The deviation is set to an explicit figure provided by the user.

Notes The deviation assigned by this technique is not the standard deviation of the Gaussians; it is the value stored in the unit threshold, which is multiplied by the distance of the weight vector from the input vector. It is related to the standard deviation by:

$$d = \frac{1}{2\sigma^2}$$

Exploratory analysis A statistical analysis that aims to discover theoretically interesting hypotheses or models describing the data.

Exploratory data analysis (EDA) As opposed to traditional hypothesis testing designed to verify a priori hypotheses about relations between variables (e.g., "There is a positive correlation between the AGE of a person and his/her RISK TAKING disposition"), exploratory data analysis (EDA) is used to identify systematic relations between variables when there are no (or not complete) a priori expectations as to the nature of those relations. In a typical exploratory data analysis process, many variables are taken into account and compared, using a variety of techniques in the search for systematic patterns.

Exponential The negative exponential function. Ideal for use with radial units. The combination of radial synaptic function and negative exponential activation function produces units that model a Gaussian (bell-shaped) function centered at the weight vector. The standard deviation of the Gaussian is given by the formula below, where d is the "deviation" of the unit stored in the unit's threshold:

$$\sigma = \sqrt{\frac{1}{d}}$$

Exponential distribution The exponential distribution function is defined as:

$f(x) = \lambda^* e^{-\lambda x}\ 0 \leq x < \infty,\ \lambda > 0$

where λ (lambda) is an exponential function parameter (an alternative parameterization is scale parameter b=1/λ)

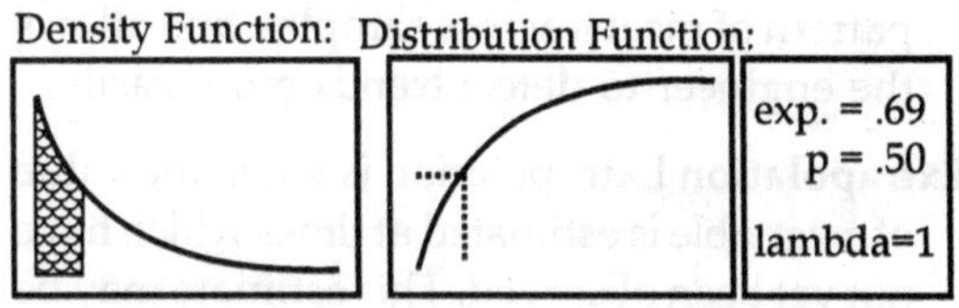

e is the base of the natural logarithm, sometimes called Euler's E.

Exponential family of distributions A family of probability distributions with exponential terms, which includes many of the most important distributions encountered in real (neural network) problems (including the normal, or Gaussian distribution, and the alpha and beta distributions).

Exponential smoothing Exponential smoothing is a smoothing technique used to reduce irregularities (random fluctuations) in time series data, thus providing a clearer view of the true underlying behaviour of the series. It also provides an effective means of predicting future values of the time series (forecasting).

Exponentially weighted moving average line This type of moving average can be considered to be a generalization of the simple moving average. Specifically, we could compute each data point for the plot as:

$z_t = \lambda^* x\text{-bar}_t + (1-\lambda)^* z_{t-1}$

In this formula, each point z_t is computed as λ (lambda) times the respective mean x- bar_t, plus one minus λtimes the previous (computed) point in the plot. The parameter λ (lambda) here should assume values greater than 0 and less than 1. You may recognize this formula as the common exponential smoothing formula. Without going into detail, this method of averaging specifies that the weight for historically "old" sample means decreases geometrically as one continues to draw samples. This type of moving average line also smoothes the pattern of means across samples, and allows the engineer to detect trends more easily.

Extrapolation Extrapolation is when the value of a variable is estimated at times which have not yet been observed. This estimate may be reasonably reliable for short times into the future, but for longer times, the estimate is liable to become less accurate.

Example

Suppose Angela was 1.20m tall on January 1st 1975, and 1.40m tall on January 1st 1976. By extrapolation, estimate her height on January 1st 1977.

By extrapolation, it could be estimated that by January 1st 1977 she would have grown another 0.20m to be 1.60m tall. This however assumes that she continued to grow at the same rate. This must eventually become a false assumption, otherwise by January 1st 1980, she would be a giantess.

Extreme event An event whose occurrence casts doubt on the null hypothesis. The idea of extremeness is relative: One event is considered to be more extreme than another if it suggests more strongly the falsity of the null hypothesis. For significance testing, extreme events are assembled into an ordered nested collection.

Extreme value distribution The extreme value (Type I) distribution (the term first used by Lieblein, 1953) has the probability density function:

$$f(x) = 1/b * e^{-(x-a)/b} * e^{-e^{-(x-a)/b}}$$

$$-\infty < x < \infty$$

$$b > 0$$

where a is the location parameter

b is the scale parameter

e is the base of the natural logarithm, sometimes called Euler's e

This distribution is also sometimes referred to as the distribution of the largest extreme.

Extreme values (in box plots) Values which are "far" from the middle of the distribution are referred to as outliers and extreme values if they meet certain conditions.

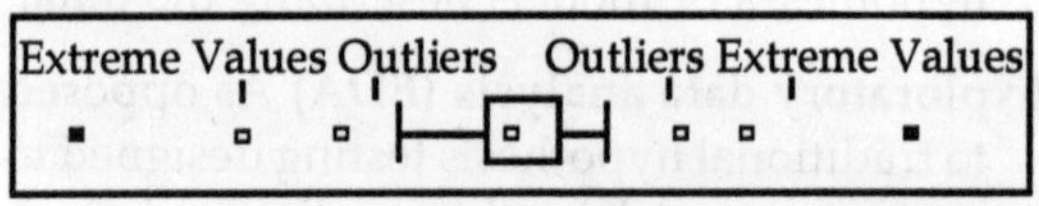

A data point is deemed to be an extreme value if the following conditions hold:

data point value > UBV + 2*o.c.*(UBV – LBV) or data point value < LBV – 2*o.c.*(UBV – LBV)

where UBV is the upper value of the box in the box plot (e.g., the mean + standard error or the 75th percentile).

LBV is the lower value of the box in the box plot (e.g., the mean - standard error or the 25th percentile).

o.c. is the outlier coefficient (when this coefficient equals 1.5, the extreme values are those which are outside the 3 box length range from the upper and lower value of the box).

F

F distribution The probability distribution of an F-ratio when the null hypothesis is true.

F ratio The ratio of two (independent) mean squares. The denominator is usually (but not always) the residual mean square. The F-ratio is typically used as a test statistic for the null hypothesis that the source of variation corresponding to the numerator in fact has no effect.

Fact Fact is a classification tree program developed by Loh and Vanichestakul (1988) that is a precursor of the quest program.

Factor A factor of an experiment is a controlled independent variable; a variable whose levels are set by the experimenter.

A factor is a general type or category of treatments. Different treatments constitute different levels of a factor. For example, three different groups of runners are subjected to different training methods. The runners are the experimental units, the training methods, the treatments, where the three types of training methods constitute three levels of the factor 'type of training'.

Factor analysis The main applications of factor analytic techniques are: 1. To reduce the number of variables and 2. To detect structure in the relationships between variables, that is to classify variables. Therefore, factor analysis is applied as a data reduction or (exploratory) structure detection method (the term factor analysis was first introduced by Thurstone, 1931).

For example, suppose we want to measure people's satisfaction with their lives. We design a satisfaction questionnaire with various items; among other things we ask our subjects how satisfied they are with their hobbies (item 1) and how intensely they are pursuing a hobby (item 2). Most likely, the responses to the two items are highly correlated with each other. Given a high correlation between the two items, we can conclude that they are quite redundant.

One can summarize the correlation between two variables in a scatterplot. A regression line can then be fitted that represents the "best" summary of the linear relationship between the variables. If we could define a variable that would approximate the regression line in such a plot, then that variable would capture most of the "essence" of the two items. Subjects' single scores on that new factor, represented by the regression line, could then be used in future data analyses to represent that essence of the two items. In a sense we have reduced the two variables to one factor.

Factorial The factorial operator is applicable to a non-negative integer quantity. It is notated as the postfixed symbol '!'. The resulting value is the product of the increasing integer values from 1 up to the value of the argument quantity. For instance : 3! is 1x2x3 = 6. By convention 0! is taken as producing the value 1. Factorial values increase very rapidly wityh increase in the argument value; this rapid growth is represented in the similarly rapid growth in numbers of combinations.

Factorial design A factorial design is used to evaluate two or more factors simultaneously. The treatments are combinations of levels of the factors. The advantages of factorial designs over one-factor-at-a-time experiments is that they are more efficient and they allow interactions to be detected.

Failure The value of a binary variable that is identified as the opposite of the event of interest, often coded as 0.

In survival analysis, a synonym for "death".

Fair bet A fair bet is one for which the expected value of the payoff is zero, after accounting for the cost of the bet. For example, suppose I offer to pay you $2 if a fair coin lands heads, but you must ante up $1 to play. Your expected payoff is –$1+ $0×P(tails) + $2×P(heads) = –$1 + $2×50% = $0. This is a fair bet—in the long run, if you made this bet over and over again, you would expect to break even.

False discovery rate In testing a collection of hypotheses, the false discovery rate is the fraction of rejected null hypotheses that are rejected erroneously (the number of Type I errors divided by the number of rejected null hypotheses), with the convention that if no hypothesis is rejected, the false discovery rate is zero.

False negative rate For a diagnostic procedure, the conditional probability of a negative result given that disease is present

False positive rate The conditional probability of a positive result given that disease is absent.

Falsification The idea that while it is impossible to prove that a hypothesis is true, it is possible to show that a hypothesis is false (since only one piece of evidence that is counter to the hypothesis is sufficient to show that it is false).Acceptance of a theory only means that we have not yet disproved it.

Feature extraction The terms feature extraction and feature selection are used in the context of predictive data mining, when the goal is to find a good predictive model for some phenomenon of interest based on a large number of predictors. While feature selection methods will attempt to identify the best predictors among the (sometimes thousands of) available predictors, feature extraction techniques attempt to aggregate or combine the predictors in some way to extract the common information contained in them that is most useful for building the model. Typical methods for feature extraction are Factor Analysis and Principal Components Analysis, Correspondence Analysis, Multidimensional Scaling, Partial Least Squares methods, or singular value decomposition, as, for example, used in text mining.

Feature selection One of preliminary stages in the process of a Data Mining applicable when the data set includes more variables than could be included (or would be efficient to include) in the actual model building phase (or even in initial exploratory operations).

Feedforward networks Neural networks with a distinct layered structure, with all connections feeding forwards from inputs towards outputs. Sometimes used as a synonym for multilayer perceptrons.

Finite population correction When sampling without replacement, as in a simple random sample, the SE of sample sums and sample means depends on the fraction of the population that is in the sample: the greater the fraction, the smaller the SE. Sampling with replacement is like sampling from an infinitely large population. The adjustment to the SE for sampling without replacement is called the finite population correction. The SE for sampling without replacement is smaller than the SE for sampling with replacement by the finite population correction factor $((N-n)/(N-1))^{½}$. Note that for sample size n=1, there is no difference between sampling with and without replacement; the finite population correction is then unity. If the sample size is the entire population of N units, there is no variability in the result of sampling without replacement (every member of the population is in the sample exactly once), and the SE should be zero. This is indeed what the finite population correction gives (the numerator vanishes).

Fisher LSD This post hoc test (or multiple comparison test) can be used to determine the significant differences between group means in an analysis of variance setting. The Fisher LSD test is considered to be one of the least conservative post hoc tests.

Fisher test(1) Named after the statistician RA Fisher. This is an exact test(1) to examine whether the pattern of counts in a 2x2 cross classification departs from expectations based upon the marginal totals for the rows and columns. Such a test is useful to examine difference in rate between two binomial outcomes. The randomisation set consists of those reassignments of the units which produce tables with the same row- and column- totals as the outcome. The randomisation set will thus consist of a number of tables with different respective patterns of counts; each such table will have a number of possible randomisations which may be a very large number. For this test there are several reasonable test statistics, including : the count in any one of the 4 cells, chi-squared(1), or the number of randomisations for each 2x2 table with the given row- and column- totals; these are equivalent test statistics. The calculation for the fisher test(1) is relatively undemanding computationally, making reference to the algebra of the hypergeometric distribution, and the test was widely used before the appearance of computers. This test has historically been regarded as superior to the use of chi-squared(2) where sample sizes are small. Statistical tables have been published for the fisher test(1) for a number of small 2x2 tables defined in terms of row- and column- totals. Also see Fisher test.

Fisher test(2) This is also known as the Freeman-Halton test. It is an extension of the logic of the fisher test(1), for a 2-way classification of counts where the extent of the cross-classification may be greater than 2x2. The randomisation set for an exhaustive randomisation test (exact test(1)) can be defined in the same way as for the fisher test(1). However, the various test statistics applicable when considering the fisher test(1) will not all be definable and will not clearly be equivalent test statistics. The test statistic which is used is the number of re-randomisations for each table with the given row- and column- totals; this test statistic has the drawback of lacking any descriptive significance in terms of the experimental design.

Fisher's exact test (for the equality of two percentages) Consider two populations of zeros and ones. Let p_1 be the proportion of ones in the first population, and let p_2 be the proportion of ones in the second population. We would like to test the null hypothesis that $p_1 = p_2$ on the basis of a simple random sample from each population. Let n_1 be the size of the sample from population 1, and let n_2 be the size of the sample from population 2. Let G be the total number of ones in both samples. If the null hypothesis be true, the two samples are like one larger sample from a single population of zeros and ones. The allocation of ones between the two samples would be expected to be proportional to the relative sizes of the samples, but would have some chance variability. Conditional on G and the two sample sizes, under the null hypothesis, the tickets in the first sample are like a random sample of size n_1 without replacement from a collection of $N = n_1 + n_2$ units of which G are labeled with ones. Thus, under the null hypothesis, the number of tickets labeled with ones in the first sample has (conditional on G) an hypergeometric distribution with parameters N, G, and n_1. Fisher's exact test uses this distribution to set the ranges of observed values of the number of ones in the first sample for which we would reject the null hypothesis.

Fitted value The value of an observation on a response variable that would be predicted by a statistical model. (Also called predicted value.) The term fitted value is used for a predicted value if the data set used in determining the prediction has an actual observed`value corresponding to the combination of explanatory variables at which the prediction is being made. With binary response variables, the term expected value is used in a similar sense.

Fitting centered polynomial models via multiple regression The fitting of higher-order polynomials of an independent variable with a mean not equal to zero can create difficult numerical problems. Specifically, the polynomials will be highly correlated due to the mean of the primary independent variable. With large numbers (e.g., Julian dates), this problem is very serious, and if proper protections are not put in place, can cause wrong results! The solution is to "center" the independent variable (sometimes, this procedures is referred to as "centered polynomials"), i.e., to subtract the mean, and then to compute the polynomials.

Fixed effects The term fixed effects in the context of analysis of variance is used to denote factors in an ANOVA design with levels that are deliberately arranged by the experimenter, rather than randomly sampled from an infinite population of possible levels (those factors are called random effects). For example, if one were interested in conducting an experiment to test the hypothesis that higher temperature leads to increased aggression, one would probably expose subjects to moderate or high temperatures and then measure subsequent aggression. Temperature would be a fixed effect in this experiment, because the levels of temperature of interest to the experimenter were deliberately set, or fixed, by the experimenter.

A simple criterion for deciding whether or not an effect in an experiment is random or fixed is to ask how one would select (or arrange) the levels for the respective factor in a replication of the study. For example, if one wanted to replicate the study described in this example, one would choose the same levels of temperature from the population of levels of temperature. Thus, the factor "temperature" in this study would be a fixed factor. If instead, one's interest is in how much of the variation of aggressiveness is due to temperature, one would probably expose subjects to a random sample of temperatures from the population of levels of different temperatures. Levels of temperature in the replication study would likely be different from the levels of temperature in the first study, thus temperature would be considered a random effect.

Football shaped scatterplot In a football-shaped scatterplot, most of the points lie within a tilted oval, shaped more-or-less like a football. A football-shaped scatterplot is one in which the data are homoscedastically scattered about a straight line.

FORTRAN A very long established and widely implemented programming language, specialised substantially for numerical applications. A number of standard programming language versions of FORTRAN have established at various dates, approved as standard by ISO.

Forward selection A procedure for exploratory analysis that successively adds more variables to the regression equation until there is no further significant improvement in the variance explained.

Frame, sampling frame A *sampling frame* is a collection of units from which a sample will be drawn. Ideally, the frame is identical to the population we want to learn about; more typically, the frame is only a subset of the population of interest. The difference between the frame and the population can be a source of bias in sampling design, if the parameter of interest has a different value for the frame than it does for the population. For example, one might desire to estimate the current annual average income of 1998 graduates of the University of California at Berkeley. I propose to use the sample mean income of a sample of graduates drawn at random. To facilitate taking the sample and contacting the graduates to obtain income information from them, I might draw names at random from the list of 1998 graduates for whom the alumni association has an accurate current address. The population is the collection of 1998 graduates; the frame is those graduates who have current addresses on file with the alumni association. If there is a tendency for graduates with higher incomes to have up-to-date addresses on file with the alumni association, that would introduce a positive bias into the annual average income estimated from the sample by the sample mean.

Free parameter A numerical value in a structural model that is part of the model, but is not fixed at any particular value by the model hypothesis. Free parameters are estimated by the program using iterative methods. Free parameters are indicated in the PATH1 language with integers placed between dashes on an arrow or a wire. For example, the following paths both have the free parameter 14.

(F1)-14->[X1]

(e1)-14-(e1)

If two different coefficients have the same free parameter number, as in the above example, then both will of necessity be assigned the same numerical value. Simple equality constraints on numerical coefficients are thus imposed by assigning them the same free parameter number.

Frequency distribution A frequency distribution is a table of frequencies of occurrence of each value of a variable. In spss 9.0 a frequency distribution is obtained by selecting Analyse, Descriptive statistics Frequencies... and usually includes a percentage calculation for each value.

Frequency histogram A graphic that displays how many measures fall into different classes, giving the frequency at which each category is seen observed.

Frequency polygon The graphical representation of the distribution of an interval or ratio variable where the midpoint of each interval is represented by a marker which is then joined by a line to the next midpoint marker.

Frequency table Frequency or one-way tables represent the simplest method for analyzing categorical (nominal) data. They are often used as one of the exploratory procedures to review how different categories of values are distributed in the sample. For example, in a survey of spectator interest in different sports, we could summarize the respondents' interest in watching football in a frequency table as follows: A frequency table is a way of summarising a set of data. It is a record of how often each value (or set of values) of the variable in question occurs. It may be enhanced by the addition of percentages that fall into each category.

A frequency table is used to summarise categorical, nominal, and ordinal data. It may also be used to summarise continuous data once the data set has been divided up into sensible groups.

When we have more than one categorical variable in our data set, a frequency table is sometimes called a contingency table because the figures found in the rows are contingent upon (dependent upon) those found in the columns.

Example

Suppose that in thirty shots at a target, a marksman makes the following scores:

5 2 2 3 4   4 3 2 0 3   0 3 2 1 5

1 3 1 5 5   2 4 0 0 4   5 4 4 5 5

The frequencies of the different scores can be summarised as

Score	Frequency	Frequency (%)
0	4	13%
1	3	10%
2	5	17%
3	5	17%
4	6	20%
5	7	23%

Frequency theory of probability See probability, theories of.

Frequentist inference An inferential paradigm that specifies certain operational rules, and provides (aleatory) probabilities about the consequences of using such rules.

Function minimization algorithms Algorithms used (e.g., in Nonlinear Estimation) to guide the search for the minimum of a function. For example, in the process of nonlinear estimation, the currently specified loss function is being minimized.

Fundamental rule of counting If a sequence of experiments or trials $T_1, T_2, T_3, \ldots, T_k$ could result, respectively, in $n_1, n_2\ n_3, \ldots, n_k$ possible outcomes, and the numbers $n_1, n_2\ n_3, \ldots, n_k$ do not depend on which outcomes actually occurred, the entire *sequence* of k experiments has $n_1 \times n_2 \times n_3 \times \ldots \times n_k$ possible outcomes.

G

G2 inverse A G2 inverse is a generalized inverse of a rectangular matrix of values A that satisfies both

$$AA'A=A \qquad \text{and} \qquad A'AA'=A$$

The g2 inverse is used to find a solution to the normal equations in the general linear model.

Gains chart The gains chart provides a visual summary of the usefulness of the

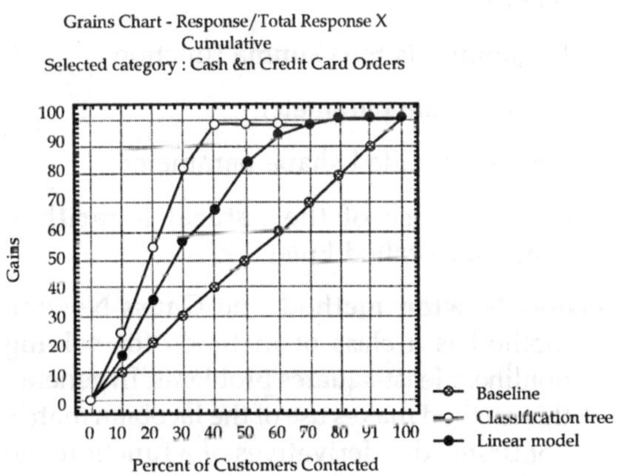

information provided by one or more statistical models for predicting a binomial (categorical) outcome variable (dependent variable); for multinomial (multiple-category) outcome variables, gains charts can be computed for each category. Specifically, the chart summarizes the utility that one can expect by using the respective predictive models, as compared to using baseline information only.

The gains chart is applicable to most statistical methods that compute predictions (predicted classifications) for binomial or multinomial responses. This and similar summary charts are commonly used in data mining projects when the dependent or outcome variable of interest is binomial or multinomial in nature.

Example. To illustrate how the gains chart is constructed, consider this example. Suppose you have a mailing list of previous customers of your business, and you want to offer to those customers an additional service by mailing an elaborate brochure and other materials describing the service. During previous similar mail-out campaigns, you collected useful information about your customers (e.g., demographic information, previous purchasing patterns) that you could relate to the response rate, i.e., whether the respective customers responded to your mail solicitation and the type of order they placed.

Given the baseline response rate and the cost of the mail-out, sending the offer to all customers would result in a net-loss. Hence, you want to use statistical analyses to help you identify the customers who are most likely to respond. Suppose you build such a model based on the data collected in the previous mail-out campaign. You can now select only the 10 percent of the customers from the mailing lists who, according to prediction from the model, are most likely to respond. Next you can compute the number of accurately predicted responses, relative to the total number of responses in the sample; this percentage is the gain due to using the model. Put another way, of those customers likely to respond in the current sample, you can accurately identify ("capture") y percent by selecting from the customer list the top 10% who were predicted by the model with the greatest certainty to respond (where y is the gains value).

Analogous values can be computed for each percentile of the population (customers on the mailing list). You could compute separate gains values for selecting the top 20% of customers who are predicted to be among likely responders to the mail campaign, the top 30%, etc. Hence, the gains values for different percentiles can be connected by a line that will typically ascend slowly and merge with the baseline if all customers (100%) were selected.

If more than one predictive model is used, multiple gains charts can be overlaid to provide a graphical summary of the utility of different models.

Game theory A field of study that bridges mathematics, statistics, economics, and psychology. It is used to study economic behavior, and to model conflict between nations, for example, "nuclear stalemate" during the Cold War.

Gamma A measure of association appropriate to variables where one or both are measured at the ordinal level of measurement.

Gamma coefficient The Gamma statistic is preferable to Spearman R or Kendall Tau when the data contain many tied observations. In terms of the underlying assumptions, Gamma is equivalent to Spearman R or Kendall tau; in terms of its interpretation and computation, it is more similar to Kendall tau than Spearman R. In short, Gamma is also a probability; specifically, it is computed as the difference between the probability that the rank ordering of the two variables agree minus the probability that they disagree, divided by 1 minus the probability of ties. Thus, Gamma is basically equivalent to Kendall tau, except that ties are explicitly taken into account.

Gamma distribution The Gamma distribution (the term first used by Weatherburn, 1946) is defined as:

$$f(x) = (x/b)^{c-1} * e^{(-x/b)} * [1/b\,\Gamma(c)]$$

$$0 \leq x,\ b > 0,\ c > 0$$

where

Γ (gamma) is the Gamma function

b is the scale parameter

a is the so-called shape parameter

e is the base of the natural logarithm, sometimes called Euler's E.

Gauss Newton method The Gauss-Newton method is a class of methods for solving nonlinear least-squares problems. In general, this method makes use of the Jacobian matrix J of first-order derivatives of a function F to find the vector of parameter values x that minimizes the residual sums of squares (sum of squared deviations of predicted values from observed values). An improved and efficient version of the method is the so-called Levenberg-Marquardt algorithm.

Gaussian distribution The normal distribution - a bell-shaped function.

General ANOVA The purpose of analysis of variance (ANOVA) is to test for significant differences between means by comparing (i.e., analyzing) variances. More specifically, by partitioning the total variation into different sources (associated with the different effects in the design), we are able to compare the variance due to the between-groups (or treatments) variability with that due to the within-group (treatment) variability. Under the null hypothesis (that there are no mean differences between groups or treatments in the population), the variance estimated from the within-group (treatment) variability should be about the same as the variance estimated from between-groups (treatments) variability.

General linear model The general linear model is a generalization of the linear regression model, such that effects can be tested 1. For categorical predictor variables, as well as for effects for continuous predictor variables and

2. In designs with multiple dependent variables as well as in designs with a single dependent variable.

Generalization The ability of a neural network to make accurate predictions when faced with data not drawn from the original training set (but drawn from the same source as the training set).

Generalized additive models Generalized Additive Models are generalizations of generalized linear models. In generalized linear models, the transformed dependent variable values are predicted from (is linked to) a linear combination of predictor variables; the transformation is referred to as the link function; also, different distributions can be assumed for the dependent variable values. An example of a generalized linear model is the Logit Regression model, where the dependent variable is assumed to be binomial, and the link function is the logit transformation. In generalized additive models, the linear function of the predictor values is replaced by an unspecified (non-parametric) function, obtained by applying a scatterplot smoother to the scatterplot of partial residuals (for the transformed dependent variable values).

Generalized inverse A generalized inverse (denoted by a superscript of -) of a rectangular matrix of values A is any matrix that satisfies

$A^{-}AA=A$

A generalized inverse of a nonsingular matrix is unique and is called the regular matrix inverse.

See also matrix singularity, matrix inverse.

Generalized linear model The generalized linear model is a generalization of the linear regression model such that 1. Nonlinear, as well as linear, effects can be tested.

2. For categorical predictor variables, as well as for continuous predictor variables, using

3. Any dependent variable whose distribution follows several special members of the exponential family of distributions (e.g., gamma, Possion, binomial, etc.), as well as for any normally-distributed dependent variable.

Generalized regression neural network (GRNN) A type of neural network using kernel-based approximation to perform regression. One of the so-called Bayesian networks.

Genetic algorithm A search algorithm which locates optimal binary strings by processing an initially random population of strings using artificial mutation, crossover and selection operators, in an analogy with the process of natural selection.

Genetic algorithm input selection Application of a genetic algorithm to determine an "optimal" set of input variables, by constructing binary masks which indicate which inputs to retain and which to discard (Goldberg, 1989). This method is implemented in Statistica Neural Networks and can be used as part of a model building process where variables identified as the most "relevant" (in statistica Neural Networks) are then used in a traditional model building stage of the analysis (e.g., using a linear regression or nonlinear estimation method).

Geometric distribution Geometric distributions model (some) discrete random variables. Typically, a Geometric random variable is the number of trials required to obtain the first failure, for example, the number of tosses of a coin untill the first 'tail' is obtained; components from a production line are tested, in turn, until the first defective item is found.

A discrete random variable X is said to follow a Geometric distribution with parameter p, written X ~ Ge(p), if it has probability distribution:

$P(X=x) = P^{x-1} (1-p)$

where:

x = 1, 2, 3, ...n

p = success probability; $0 < p < 1$

The trials must meet the following requirements:

a. The total number of trials is potentially infinite;

b. There are just two outcomes of each trial; success and failure;

c. The outcomes of all the trials are statistically independent;

d. All the trials have the same probability of success.

The Geometric distribution has expected value E(X):

$$E(X) = \frac{1}{1-p}$$

and variance V(X):

$$V(X) = \frac{p}{(1-p)^2}$$

The Geometric distribution is related to the Binomial distribution in that both are based on independent trials in which the probability of success is constant and equal to p. However, a Geometric random variable is the number of trials until the first failure, whereas a Binomial random variable is the number of successes in n trials.

Geometric mean The geometric mean of *n* numbers $\{x_1, x_2, x_3, \ldots, x_n\}$ is the *n*th root of their product:

$(x_1 \times x_2 \times x_3 \times \ldots \times x_n)^{1/n}$.

Gibbs sampler The Gibbs sampler is a popular method used for MCMC (markov chain monte carlo) analyses. It provides an elegant way for sampling from the joint distributions of multiple variables, by applying the notion that: to sample from a joint distribution just sample repeatedly from its one-dimensional conditionals given whatever you've seen at the time.

For example, the values from the joint distribution of two random variables, X and Y, can be easily simulated by the Gibbs sampler that uses their conditional distributions rather than their joint distribution. Starting with an arbitrary choice of X and Y, X is simulated from the conditional distribution of X, given Y, and Y is simulated from conditional distribution of Y, given X. Alternating between two conditional distributions, in the subsequent steps, generates a sample from the correct joint distribution of X and Y; the approximation gets better and better as the length of the Gibbs sampler path increases.

Gini measure of node impurity According to Breiman, Friedman, Olshen, & Stone (1984), the Gini measure of node impurity at node (which statistica uses by default in GC&RT and, therefore, Boosted Trees) is defined to be

$$\text{impurity}(t) = \sum_{i-j} p(i|t)p(j|t)$$

where

$$p(j \mid t) = \frac{p(j,t)}{p(t)}$$

and

$$p(j,t) = \frac{\pi(j)N_i(t)}{N_i}$$

such that

p (j t) is the estimated probability that an observation belongs to group j given that it is in node t,

p (j , t) is the estimated probability that an observation is in group j and at node t,

p (t) is the estimated probability that an observation is at node t, $\frac{N(t)}{N}$,

$\pi(j)$ is the prior probability for group j,

N j (t) is the number of group j members at node t,

and N_j is the size of group j.

Therefore, the prior probabilities play a role in every Gini Measure computation at every node. However, Breiman et al. also note that, when the prior probabilities are estimated from the data,

$$\pi(j) = \frac{N_j}{N} \Rightarrow p(j|t) = \frac{N_j(t)}{N(t)}$$

This fact can cause higher misclassification rates in under-represented groups.

Gold standard(1) The gold standard is the form of test which is most faithful to the randomisation distribution, for a given test statistic and experimental design. This involves exhaustive randomisation. Other randomisation tests may reasonably be judged by comparison with this form.

Gold standard(2) The idea of a re-randomisation test as a standard of correctness by which to judge other tests which are not based upon principles of re-randomisation.

Gompertz distribution The Gompertz distribution is a theoretical distribution of survival times. Gompertz (1825) proposed a probability model for human mortality, based on the assumption that the "average exhaustion of a man's power to avoid death to be such that at the end of equal infinetely small intervals of time he lost equal portions of his remaining power to oppose destruction which he had at the commencement of these intervals".

Goodman and Kruskal's TAU An asymmetric measure of association between two variables appropriate when one variable is considered to be the cause and the other the effect.

Goodness of fit Various goodness-of-fit summary statistics can be computed for continuous and categorical dependent variables. Most of these statistics are discussed in greater detail in Witten and Frank (2000); in the context of forecasting; different statistics are also discussed in Makridakis and Wheelwright (1983). Goodness of fit statistics for regression problems (for continuous variables) include:

least squares deviation (LSD), mean square error

Average deviation, mean absolute error

Relative squared error, mean relative squared error

Correlation coefficient (Pearson product moment correlation)

Goodness of fit statistics for classification problems (for categorical variables) include:

Pearson Chi-square

G-square (maximum likelihood Chi-square)

Percent disagreement (misclassification rate).

Gradient In Structural Equation Modeling the gradient is the vector of first partial derivatives of the discrepancy function with respect to the parameter values. At a local or global minimum, the discrepancy function should be at the bottom of a "valley," where all first partial derivatives are zero, so the elements of the gradient should all be near to zero when a minimum is obtained.

The elements of the gradient, by themselves, can, on occasion, be somewhat unreliable as indicators of when convergence has occurred, especially when the model fit is not good, and the discrepancy function value itself is quite large. For this reason, the gradient is not employed as a convergence criterion by this program.

Gradient descent Optimization techniques for non-linear functions (e.g. the error function of a neural network as the weights are varied) which attempt to move incrementally to successively lower points in search space, in order to locate a minimum.

Gradual permanent impact In Time Series, the gradual permanent impact pattern implies that the increase or decrease due to the intervention is gradual, and that the final permanent impact becomes evident only after some time. This type of intervention can be summarized by the expression:

$\text{Impact}_t = \delta^* \text{Impact}_{t-1} + \omega$

(for all t ≥time of impact, else = 0).

Note that this impact pattern is defined by the two parameters δ (delta) and ω (omega). If δ is near 0 (zero), then the final permanent amount of impact will be evident after only a few more observations; if δ is close to 1, then the final permanent amount of impact will only be evident after many more observations. As long as the d parameter is greater than 0 and less than 1 (the bounds of system stability), the impact will be gradual and result in an asymptotic change (shift) in the overall mean by the quantity:

Asymptotic change in level = ω/(1-δ)

Graph of averages For bivariate data, a graph of averages is a plot of the average values of one variable (say *y*) for small ranges of values of the other variable (say *x*), against the value of the second variable (*x*) at the midpoints of the ranges.

Group charts See multiple stream group charts.

Grouped data Data which has been recoded or grouped into fewer categories than when originally collected. For instance, age, measured on an interval scale, could be grouped into agegroups in order to create a histogram.

Grouped measures A set of values that belong to the same class.

Grouping (or coding) variable A grouping (or coding) variable is used to identify group membership for individual cases in the data file. Typically, the grouping variable is categorical (i.e., contains either discrete values, e.g., 1, 2, 3,

or a few text values, e.g., male, female)

and the values are referred to as codes (they can be integer values or integer values with text value equivalents).

Groupware Software intended to enable a group of users on a network to collaborate on specific projects. Groupware can provide services for communication (such as e-mail), collaborative document development, analysis, reporting, statistical data analysis, scheduling, or tracking. Documents can include text, images, or any other forms of information (e.g., multimedia).

H

H

Half normal probability plots. This type of graph is used to evaluate the normality of the distribution of a variable, that is, whether and to what extent the distribution of the variable follows the normal distribution. The selected variable will be plotted in a scatterplot against the values "expected from the normal distribution." The half-normal probability plot is constructed in the same way as the standard normal probability plot, except that only the positive half of the normal curve is considered. Consequently, only positive normal values will be plotted on the Y-axis. This plot is used when one wants to ignore the sign of the residual, that is, when one is mostly interested in the distribution of absolute residuals, regardless of the sign.

Hamming window In Time Series, the Hamming window is a weighted moving average transformation used to smooth the periodogram values. In the Hamming (named after R. W. Hamming) window or Tukey- Hamming window (Blackman and Tukey, 1958), for each frequency, the weights for the weighted moving average of the periodogram values are computed as:

$w_j = 0.54 + 0.46*\text{cosine}(\pi*j/p)$ (for j=0 to p)

$w_{-j} = w_j$ (for $j \neq 0$)

where p = (m-1)/2

This weight function will assign the greatest weight to the observation being smoothed in the center of the window, and increasingly smaller weights to values that are further away from the center.

Hanging bars histogram The hanging bars histogram offers a "visual test of normality"

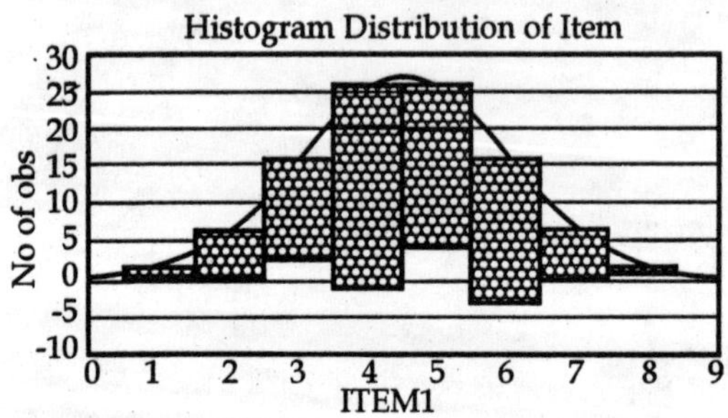

of the distribution that helps identify the areas of the distribution where the discrepancies (between the observed and expected normal frequencies) occur. While the standard way of presenting the normal distribution fitted to an observed distribution is to overlay the best-fitting normal curve over a histogram, the hanging bars histogram does just the opposite: it suspends the bars representing the observed frequencies for consecutive ranges of values from the best-fitting normal curve.

If the investigated distribution can be well approximated by the normal curve, then the bottoms of all bars should form a straight, horizontal line.

Harmonic mean The Harmonic Mean is a "summary" statistic used in analyses of frequency data; it is computed as:

$$H = n * 1/\Sigma (1/x_i)$$

where n is the sample size.

Hazard It is often meaningful to consider the function that describes the probability of failure during a very small time increment (assuming that no failures have occurred prior to that time). This function is called the hazard function (or, sometimes, also conditional failure, intensity, or force of mortality function), and is generally defined as:

$$h(t) = f(t)/(1-F(t))$$

where h(t) stands for the hazard function (of time t), and f(t) and F(t) are the probability density and cumulative distribution functions, respectively.

Hazard function (often called just hazard.) For an individual at a given age, the limiting ratio (as the length of the interval shrinks to zero) of the probability of dying in the next interal of time divided by the length of the interval.

Hazard rate In Survival Analysis the hazard rate is defined as the probability per time unit that a case that has survived to the beginning of the respective interval will fail in that interval. Specifically, it is computed as the number of failures per time units in the respective interval, divided by the average number of surviving cases at the mid-point of the interval.

Heteroscedasticity "Mixed scatter." A scatterplot or residual plot shows heteroscedasticity if the scatter in vertical slices through the plot depends on where you take the slice. Linear regression is not usually a good idea if the data are heteroscedastic.

Heuristic As opposed to an algorithm (which contains a fully defined set of steps that will produce a specific outcome), heuristics are general recommendations or guides based on statistical evidence (e.g., "quit smoking to prolong your life," "males with college education are more likely to respond positively to this advertisement than...") or theoretical reasoning (e.g., "the mechanism of the vitamin X synthesis as we understand it, implies that eating Y will reduce the deficit of X").

Heywood case A Heywood case in common factor analysis occurs when the minimum of the discrepancy function is obtained with one or more negative values as estimates for the variance of the unique variables. Such values are of course impossible. Heywood cases occur frequently when too many factors are extracted, or the sample size is too small.

Hidden layers All layers of a neural network except the input and output layers. Hidden layers provide the network's non-linear modeling capabilities.

High low close In this type of box or range plot, the "serifs" on the whiskers are not symmetrical but point to the left of the bar, representing the traditional "stock pricc graph" style. Note that you can change the whisker style (i.e., Hi/Lo Left, Hi/Lo Right, or Whiskers), for example.

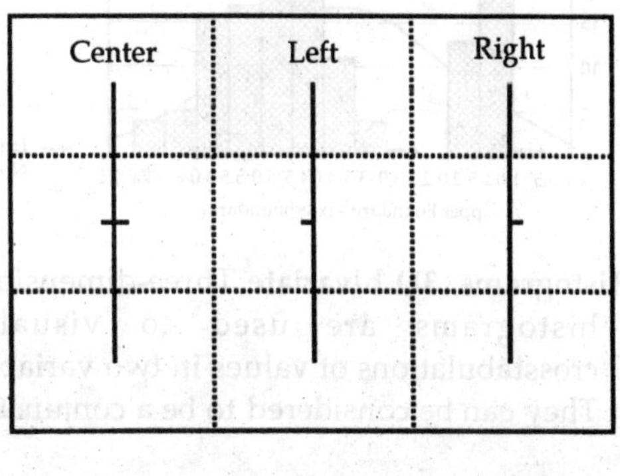

Histogram A histogram is a kind of plot that summarizes how data are distributed. Starting with a set of class intervals, the histogram is a set of rectangles ("bins") sitting on the horizontal axis. The bases of the rectangles are the class intervals, and their heights are such that their areas are proportional to the fraction of observations in the corresponding class intervals. That is, the height of a given rectangle is the fraction of observations in the corresponding class interval, divided by the length of the corresponding class interval. A histogram does not need a vertical scale, because the total area of the histogram must equal 100%. The units of the vertical axis are percent per unit of the horizontal axis. This is called the *density scale*. The horizontal axis of a histogram needs a scale. If any observations coincide with the endpoints of class intervals, the endpoint convention is important.

Histograms, 2D 2D histograms (the term was first used by Pearson, 1895) present a graphical representation of the frequency distribution of the selected variable(s) in which the columns are drawn over the class intervals and the heights of the columns are proportional to the class frequencies.

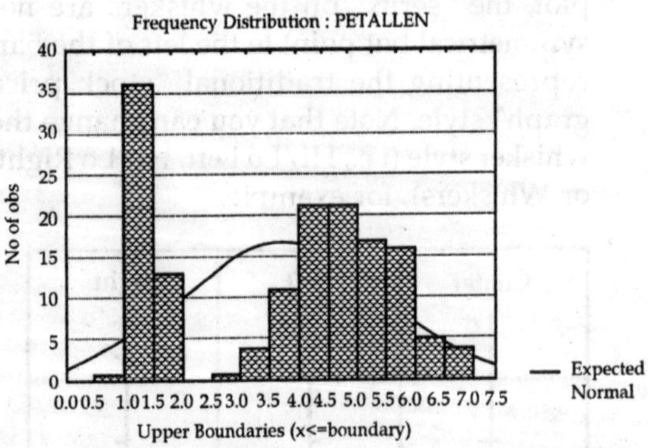

Histograms, 3D bivariate Three-dimensional histograms are used to visualize crosstabulations of values in two variables. They can be considered to be a conjunction of two simple (i.e., univariate) histograms, combined such that the frequencies of co-occurrences of values on the two analyzed variables can be examined. In a most common format of this graph, a 3D bar is drawn for each "cell" of the crosstabulation table and the height of the bar represents the frequency of values for the respective cell of the table. Different methods of categorization can be used for each of the two variables for which the bivariate distribution is visualized.

Histograms, 3D box plots This type of bivariate histogram represents the frequencies as a series of 3D bars ("rectangular boxes"). This is the default representation of 3D histograms. The "height" of each bar on the Z- axis corresponds to the frequency of the respective combination of levels for the two variables.

Histograms, 3D contour plot This contour plot presents a projection of the spline-smoothed surface fit to the frequency data. Successive values of each series are plotted along the X-axis, with each successive series represented along the Y-axis.

Histograms, 3D spikes In this type of bivariate histogram, the frequencies are represented as a series of "spikes" (point symbols with lines descending to the base plane). The "height" of each spike is determined by the frequency for the respective combination of levels of the two variables.

Histograms, 3D surface plot In this representaion of the 3D bivariate histogram, a spline-smoothed surface is fit to the frequency data.

Historical controls Sometimes, the a treatment group is compared with individuals from another epoch who did not receive the treatment; for example, in studying the possible effect of fluoridated water on childhood cancer, we might compare cancer rates in a community before and after fluorine was added to the water supply. Those individuals who were children before fluoridation started would comprise an historical control group. Experiments and studies with historical controls tend to be more susceptible to confounding than those with contemporary controls, because many factors that might affect the outcome other than the treatment tend to change over time as well. (In this example, the level of other potential carcinogens in the environment also could have changed.)

Hollander Proschan test This test compares the theoretical reliability function to the Kaplan-Meier estimate. The actual computations for this test are somewhat complex. The Hollander-Proschan test is applicable to complete, single-censored, and multiple-censored data sets; however, Dodson (1994) cautions that the test may sometimes indicate a poor fit when the data are heavily single-censored. The Hollander-Proschan C statistic can be tested against the normal distribution (z).

The Hollander-Proschan test is used in Weibull and Reliability/Failure Time Analysis.

Homoscedasticity Same scatter. A scatterplot or residual plot shows homoscedasticity if the scatter in vertical slices through the plot does not depend much on where you take the slice.

Hooke Jeeves pattern moves A Nonlinear Estimation procedure which at each iteration, first defines a pattern of points by moving each parameter one by one, so as to optimize the current loss function. The entire pattern of points is then shifted or moved to a new location; this new location is determined by extrapolating the line from the old base point in the m dimensional parameter space to the new base point. The step sizes in this process are constantly adjusted to "zero in" on the respective optimum. This method is usually quite effective, and should be tried if both the quasi-Newton and Simplex methods fail to produce reasonable estimates.

House edge In casino games, the expected payoff to the bettor is negative: the house (casino) tends to win money in the long run. The amount of money the house would expect to win for each $1 wagered on a particular bet (such as a bet on "red" in roulette) is called the *house edge* for that bet.

HTML Acronym for HyperText Markup Language. The markup language used for documents on the World Wide Web. HTML uses tags to identify elements of the document, such as text or graphics. HTML 2.0, defined by the Internet Engineering Task Force (IETF), includes features of HTML common to all Web browsers as of 1995 and was the first version of HTML widely used on the World Wide Web. Future HTML development will be carried out by the World Wide Web Consortium (W3C). HTML 3.2, the latest proposed standard, incorporates features widely implemented as of early 1996. Most Web browsers, notably Netscape Navigator and Internet Explorer, recognize HTML tags beyond those included in the present standard.

Hyperbolic The hyperbolic tangent function (tanh): a sigmoid curve, like the logistic function, except that output lies in the range (-1,+1). Often performs better than the logistic function because of its symmetry. Ideal for customization of multilayer perceptrons, particularly the hidden layers.

Hyperbolic tangent (TANH) A symmetric S-shaped (sigmoid) function, sometimes used as an alternative to logistic functions.

Hypergeometric distribution The hypergeometric distribution with parameters *N*, *G* and *n* is the distribution of the number of "good" objects in a simple random sample of size *n* (*i.e.*, a random sample without replacement in which every subset of size n has the same chance of occurring) from a population of N objects of which G are "good." The chance of getting exactly g good objects in such a sample is

$${}_{G}C_{g} \times {}_{N-G}C_{n-g} / {}_{N}C_{n},$$

provided $g <= n$, $g <= G$, and $n - g <= N - G$. (The probability is zero otherwise.) The expected value of the hypergeometric distribution is $n \times G/N$, and its standard error is

$$((N-n)/(N-1))^{½} \times (n \times G/N \times (1-G/N))^{½}.$$

Hyperplane An N-dimensional analogy of a line or plane, which divides an N+1 dimensional space into two.

Hypersphere An N-dimensional analogy of a circle or sphere.

Hypothesis test Setting up and testing hypotheses is an essential part of statistical inference. In order to formulate such a test, usually some theory has been put forward, either because it is believed to be true or because it is to be used as a basis for argument, but has not been proved, for example, claiming that a new drug is better than the current drug for treatment of the same symptoms.

In each problem considered, the question of interest is simplified into two competing claims / hypotheses between which we have a choice; the null hypothesis, denoted H0, against the alternative hypothesis, denoted H1. These two competing claims / hypotheses are not however treated on an equal basis, special consideration is given to the null hypothesis. We have two common situations:

1. The experiment has been carried out in an attempt to disprove or reject a particular hypothesis, the null hypothesis, thus we give that one priority so it cannot be rejected unless the evidence against it is sufficiently strong. For example, H0: there is no difference in taste between coke and diet coke against H1: there is a difference.

2. If one of the two hypotheses is 'simpler' we give it priority so that a more 'complicated' theory is not adopted unless there is sufficient evidence against the simpler one. Statistical hypothesis testing is formalized as making a decision between rejecting or not rejecting a null hypothesis, on the basis of a set of observations. Two types of errors can result from any decision rule (test): rejecting the null hypothesis when it is true (a Type I error), and failing to reject the null hypothesis when it is false (a Type II error). For any hypothesis, it is possible to develop many different decision rules (tests). Typically, one specifies ahead of time the chance of a Type I error one is willing to allow. That chance is called the significance level of the test or decision rule. For a given significance level, one way of deciding which decision rule is best is to pick the one that has the smallest chance of a Type II error when a given alternative hypothesis is true. The chance of correctly rejecting the null hypothesis when a given alternative hypothesis is true is called the power of the test against that alternative.

I

I.I.D An abbreviation for "independent and identically distributed", used adjectivally about sequences of random variables.

Icon plots One of the potentially powerful general techniques of exploratory data analysis are multidimensional icon graphs. The basic idea of icon plots is to represent individual units of observation as particular graphical objects where values of variables are assigned to specific features or dimensions of the objects (usually one case = one object). The assignment is such that the overall appearance of the objects changes as a function of the configuration of values. Thus, the objects are given visual "identities" that are unique for configurations of values and that can be identified by the observer. Examining such icons may help to discover specific clusters of both simple relations and interactions between variables.

Icon plots (Chernoff faces) Chernoff faces is the most "elaborate" type of icon plot. A separate "face" icon is drawn for each case; relative values of the selected variables for each case are assigned to shapes and sizes of individual facial features (e.g., length of nose, angle of eyebrows, width of face).

Icon plots columns In this type of icon plot, an individual column graph is plotted for each case; relative values of the selected variables for each case are represented by the height of consecutive columns.

Icon plots lines In this type of icon plot, an individual line graph is plotted for each case; relative values of the selected variables for each case are represented by the height of consecutive break points of the line above the baseline.

Icon plots pies In this type of icon plot, data values for each case are plotted as a pie chart (clockwise, starting at 12:00); relative values of selected variables are represented by the size of the pie slices.

Icon plots polygons In this type of icon plot, a separate polygon icon is plotted for each case; relative values of the selected variables for each case are represented by the distance from the center of the icon to consecutive corners of the polygon (clockwise, starting at 12:00).

Icon plots profiles In this type of icon plot, an individual area graph is plotted for each case; relative values of the selected variables for each case are represented by the height of consecutive peaks of the profile above the baseline.

Icon plots stars In this type of icon plot, a separate star-like icon is plotted for each case; relative values of the selected variables for each case are represented (clockwise, starting at 12:00) by the relative length of individual rays in each star. The ends of the rays are connected by a line.

Icon plots sun rays In this type of icon plot, a separate sun-like icon is plotted for each case; each ray represents one of the selected variables (clockwise, starting at 12:00), and the length of the ray represents 4 standard deviations. Data values of the variables for each case are connected by a line.

Identity The activation level is passed on directly as the output. Used in a variety of network types, including linear networks, and the output layer of radial basis function networks.

IFF (if and only if) If *p* and *q* are two logical propositions, then(*p* IFF *q*) is a proposition that is true when both *p* and *q* are true, and when both *p* and *q* are false. If is logically equivalent to the proposition

((*p* implies *q*) and (*q* implies *p*)) and to the proposition

((*p* and *q*) or ((not *p*) and (not *q*))).

Illustrations a. The more the points tend to cluster around a straight line, the stronger the linear relationship between the two variables (the higher the correlation).

b. If the line around which the points tends to cluster runs from lower left to upper right, the relationship between the two variables is positive (direct).

c. If the line around which the points tends to cluster runs from upper left to lower right, the relationship between the two variables is negative (inverse).

d. If there exists a random scatter of points, there is no relationship between the two variables (very low or zero correlation).

e. Very low or zero correlation could result from a non-linear relationship between the variables. If the relationship is in fact non-linear (points clustering around a curve, not a straight line), the correlation coefficient will not be a good measure of the strength.

A scatterplot will also show up a non-linear relationship between the two variables and whether or not there exist any outliers in the data.

More information can be added to a two-dimensional scatterplot - for example, we might label points with a code to indicate the level of a third variable.

If we are dealing with many variables in a data set, a way of presenting all possible scatter plots of two variables at a time is in a scatterplot matrix.

Implies, logical implication Logical implication is an operation on two logical propositions. If *p* and *q* are two logical propositions, (*p* implies *q*) is a logical proposition that is true if *p* is false, or if both *p* and *q* are true. The proposition (*p* implies *q*) is logically equivalent to the proposition ((not *p*) or *q*).

Impossible event An event that never happens, commonly denoted by ". (This is a circle with a slash - some browsers may not display this.)

Improve The goal of the Improve phase is to implement solutions that address the problems (root causes) identified during the previous (Analyze) phase.

In place database processing (IDP) In place database processing (IDP) is an advanced database access technology developed at StatSoft to support high-performance, direct interface between external data sets residing on remote servers and the analytic functionality of data analysis software (such as Statistica products) residing on the client computers. The IDP technology has been developed to facilitate accessing data in large databases using a one-step process which does not necessitate creating local copies of the data set. IDP significantly increases the overall performance of data processing software; it is particularly well suited for large data mining and exploratory data analysis tasks.

The source of the IDP performance gains. The speed gains of the IDP technology - over accessing data in a traditional way - result not only from the fact that IDP allows the client data analysis software to access data directly in databases and skip the otherwise necessary step of first importing the data and creating a local data file, but also from its "multitasking" (technically, asynchronous and distributed processing) architecture. Specifically, IDP uses the processing resources (multiple CPUs) of the database server computers to execute the query operations, extract the requested records of data and send them to the client computer, while the data analysis software on the client computer is simultaneously processing these records as they arrive.

Incremental (vs. non-incremental learning) algorithms Methods (algorithms) for predictive data mining are also referred to as "learning" algorithms, because they derive information from the data to predict new observations. These algorithms can be divided into those that require one or perhaps two complete passes through the input data, and those that require iterative multiple access to the data to complete the estimation. The former type of algorithms are also sometimes referred to as incremental learning algorithms, because they will complete the computations necessary to fit the respective models by processing one case at a time, each time "refining" the solution; then, when all cases have been processed, only few additional computations are necessary to produce the final results. Non-incremental learning algorithms are those that need to process all observations in each iteration of an iterative procedure for refining a final solution. Obviously, incremental learning algorithms are usually much faster than non-incremental algorithms, and for extremely large data sets, non-incremental algorithms may not be applicable at all (without first sub-sampling).

Independent Of two events, the property that the probability of their intersection is the product of their probabilities; of a collection of more than two events, the property that the probability of the intersection of any subcollection is the product of their respective probabilities. Intuitively, one thinks of a collection of events as being independent if there is no information about any one of the events contained in any combination of the others.

Random variables in a collection are said to be independent if any event described by one of the random variables is independent of every event described by any combination of the others.

Independent and identically distributed (IID) A collection of two or more random variables $\{X_1, X_2, \ldots, \}$ is *independent and identically distributed* if the variables have the same probability distribution, and are independent.

Independent events Two events are independent if the occurrence of one of the events gives us no information about whether or not the other event will occur; that is, the events have no influence on each other.

In probability theory we say that two events, A and B, are independent if the probability that they both occur is equal to the product of the probabilities of the two individual events. That is:

$$P(A \cap B) = P(A).P(B)$$

The idea of independence can be extended to more than two events. For example, A, B and C are independent if:

a. A and B are independent; A and C are independent and B and C are independent (pairwise independence);

b. $P(A \cap B \cap C) = P(A).P(B).P(C)$

If two events are independent then they cannot be mutually exclusive (disjoint) and vice versa.

Example

Suppose that a man and a woman each have a pack of 52 playing cards. Each draws a card from his/her pack. Find the probability that they each draw the ace of clubs.

We define the events:

A = probability that man draws ace of clubs = 1/52

B = probability that woman draws ace of clubs = 1/52

Clearly events A and B are independent so:

$$P(A \cap B) = P(A).P(B) = \frac{1}{52}.\frac{1}{52} = 0.00037$$

That is, there is a very small chance that the man and the woman will both draw the ace of clubs.

See also conditional probability.

Independent random variables Two random variables X and Y say, are said to be independent if and only if the value of X has no influence on the value of Y and vice versa.

The cumulative distribution functions of two independent random variables X and Y are related by

$F(x,y) = G(x).H(y)$

where:

G(x) and H(y) are the marginal distribution functions of X and Y for all pairs (x,y).

Knowledge of the value of X does not effect the probability distribution of Y and vice versa. Thus there is no relationship between the values of independent random variables.

For continuous independent random variables, their probability density functions are related by

$f(x,y) = g(x).h(y)$

where:

g(x) and h(y) are the marginal density functions of the random variables X and Y respectively, for all pairs (x,y).

For discrete independent random variables, their probabilities are related by

$P(X = xi ; Y = yj) = P(X = xi).P(Y=yj)$

for each pair (xi,yj).

Independent sampling Independent samples are those samples selected from the same population, or different populations, which have no effect on one another. That is, no correlation exists between the samples.

Independent variable In regression, the independent variable is the one that is supposed to explain the other; the term is a synonym for "explanatory variable." Usually, one regresses the "dependent variable" on the "independent variable." There is not always a clear choice of the independent variable. The independent variable is usually plotted on the horizontal axis. Independent in this context does not mean the same thing as statistically independent.

Independent vs. dependent variables The terms dependent and independent variable apply mostly to experimental research where some variables are manipulated, and in this sense they are "independent" from the initial reaction patterns, features, intentions, etc. of the subjects. Some other variables are expected to be "dependent" on the manipulation or experimental conditions. That is to say, they depend on "what the subject will do" in response. Independent variables are those that are manipulated whereas dependent variables are only measured or registered.

Somewhat contrary to the nature of this distinction, these terms are also used in studies where we do not literally manipulate independent variables, but only assign subjects to "experimental groups" based on some preexisting properties of the subjects. For example, if in an experiment, males are compared with females regarding their white cell count (WCC), Gender could be called the independent variable and WCC the dependent variable.

Independence Two events A and B are (statistically) independent if the chance that they both happen simultaneously is the product of the chances that each occurs individually; *i.e.*, if P(AB) = P(A)P(B). This is essentially equivalent to saying that learning that one event occurs does not give any information about whether the other event occurred too: the conditional probability of A given B is the same as the unconditional probability of A, *i.e.*, P(A | B) = P(A). Two random variables X and Y are independent if all events they determine are independent, for example, if the event $\{a < X <= b\}$ is independent of the event $\{c < Y <= d\}$ for **all** choices of *a*, *b*, *c*, and *d*. A collection of more than two random variables is independent if for every proper subset of the variables, every event determined by that subset of the variables is independent of every event determined by the variables in the complement of the subset. For example, the three random variables X, Y, and Z are independent if every event determined by X is independent of every event determined by Y and every event determined by X is independent of every event determined by Y and Z and every event determined by Y is independent of every event determined by X and Z and every event determined by Z is independent of every event determined by X and Y.

Indicator A method intended to measure a concept. For instance, a common indicator of the concept of social class is a person's occupation. Since class is not something we can measure directly, an indicator has to be used.

Indicator random variable The indicator [random variable] of the event A, often written 1_A, is the random variable that equals unity if A occurs, and zero if A does not occur. The expected value of the indicator of A is the probability of A, P(A), and the standard error of the indicator of A is $(P(A)\times(1-P(A))^{1/2}$. The sum

$$1_A + 1_B + 1_C + \cdot$$

of the indicators of a collection of events {A, B, C, . . .} counts how many of the events {A, B, C, . . .} occur in a given trial. The product of the indicators of a collection of events is the indicator of the intersection of the events (the product equals one if and only if all of indicators equal one). The maximum of the indicators of a collection of events is the indicator of the union of the events (the maximum equals one if any of the indicators equals one).

Indicator variable A binary variable that is coded with the values 0 and 1. A 1 represents the presence of some characteristic, whereas 0 represents its absence.

Individual (also called *unit, case, observation,* or *subject.*) A member of a population.

Induction A method of analysis that derives theories by generalising from evidence, usually from a large number of cases.

Inertia The term inertia in correspondence analysis is used by analogy with the definition in applied mathematics of "moment of inertia," which stands for the integral of mass times the squared distance to the centroid. Inertia is defined as the total Pearson Chi-square for a two-way frequency table divided by the total sum of all observations in the table.

Inference Conclusion about a population parameter based upon analysis of a sample statistic. Inferences are always stated with a confidence level.

Inferential statistics Statistics used to determine whether the changes observed in a sample are likely to be true of the population which the sample is intended to represent.

Inner fence In a boxplot, the inner fences are the boundaries of the main body of the data beyond which lie the outliers. It is positioned at 1.5 times the IQR above or below the upper and lower quartiles respectively.

Inter quartile range (IQR) The inter-quartile range is a measure of the spread of or dispersion within a data set.

It is calculated by taking the difference between the upper and the lower quartiles. For example:

Data 2 3 4 5 6 6 6 7 7 8 9

Upper quartile 7

Lower quartile 4

IQR 7 - 4 = 3

The IQR is the width of an interval which contains the middle 50% of the sample, so it is smaller than the range and its value is less affected by outliers.

Interactions An effect of interaction occurs when a relation between (at least) two variables is modified by (at least one) other variable. In other words, the strength or the sign (direction) of a relation between (at least) two variables is different depending on the value (level) of some other variable(s). (The term interaction was first used by Fisher, 1926). Note that the term "modified" in this context does not imply causality but represents a simple fact that depending on what subset of observations (regarding the "modifier" variable(s)) you are looking at, the relation between the other variables will be different.

For example, imagine that we have a sample of highly achievement-oriented students and another of achievement "avoiders." We now create two random halves in each sample, and give one half of each sample a challenging test, the other an easy test. We measure how hard the students work on the test. The means of this (fictitious) study are as follows:

Achievement-oriented
Achievement-avoiders

Challenging TestEasy Test 10 5 510

s (fictitious) study are as follows:

How can we summarize these results? Is it appropriate to conclude that 1. Challenging tests make students work harder,

2. Achievement-oriented students work harder than achievement-avoiders? None of these statements captures the essence of this clearly systematic pattern of means. The appropriate way to summarize the result would be to say that challenging tests make only achievement-oriented students work harder, while easy tests make only achievement-avoiders work harder. In other words, the relation between the type of test and effort is positive in one group but negative in the other group. Thus, the type of achievement orientation and test difficulty interact in their effect on effort; specifically, this is an example of a two-way interaction between achievement orientation and test difficulty. (Note that statements 1 and 2 above would describe so-called main effects.).

Intercept The value of y at which a line crosses the vertical axis.

Interpolation Given a set of bivariate data (x, y), to impute a value of y corresponding to some value of x at which there is no measurement of y is called interpolation, if the value of x is within the range of the measured values of x. If the value of x is outside the range of measured values, imputing a corresponding value of y is called extrapolation.

Interpreter A program supplied especially for a particular type of computer, to enable the translation of code expressed in some programming language into object code for that type of computer. An interpreter undertakes translation of the user's program in small functional units (statements) to object code as the program is used and allows modification of the sequence of statements without need to generate a full explicit object code version of the program; this is in contrast to the action of a compiler. Use of an interpreter is convenient and flexible for program development; however, running a program produced in this way generally requires more computational resource (particularly in terms of run time) than for the object code produced using a compiler.

Intersection The intersection of two or more sets is the set of elements that all the sets have in common; the elements contained in every one of the sets. The intersection of the events A and B is written "A and B" and "AB."

Interval A scale using numbers to rank order; its intervals are equal but with an arbitrary 0 point.

Interval estimator An estimator providing a range of values as estimates. cf. point estimator.

Interval scale A characteristic of data such that the difference between two values measured on the scale has the same substantive meaning/significance irrespective of the common level of the two values being compared. This implies that scores may meaningfully be added or subtracted and that the mean is a representative measure of central tendency. Such data are common in the domain of physical sciences or engineering - e.g. lengths or weights.

Interval variable An interval variable is one which categories may not only be ranked or ordered but the distance between the categories is precisely defined e.g. salary, age. See level of measurement

Intervening A variable is intervening if another variable has an effect on it and it in turn affects a third variable.

Intraclass correlation coefficient The value of the population intraclass correlation coefficient is a measure of the homogeneity of observations within the classes of a random factor relative to the variability of such observations between classes. It will be zero only when the estimated effect of the random factor is zero and will reach unity only when the estimated effect of error is zero, given that the the total variation of the observations is greater than zero.

Note that the population intraclass correlation can be estimated using variance component estimation methods.

Invariance under a constant scale factor (ICSF) A structural model is invariant under a constant scale factor (ICSF) if model fit is not changed if all variables are multiplied by the same constant. Most, but not all, structural models that are of practical interest are ICSF (see structural equation modeling).

Invariance under change of scale (ICS) A structural model is invariant under change of scale if model fit is not changed by rescaling the variables, i.e., by multiplying them by scale factors (see structural equation modeling).

Inverse document frequency The inverse document frequency is a useful transformation of raw word frequency counts computed in the context of text mining, in order to simultaneously express the frequencies with which specific terms or words are used in a collection of documents, as well as their semantic specificities, i.e., the extent to which particular words are used only in specific documents in the collection.

Suppose you index a collection of text documents and compute the word frequencies (wf) to enumerate the number of times that each word or term is used in each document. A particular issue that you may want to consider more carefully, and reflect in the indices used in further analyses, are the relative document frequencies (df) of different words. For example, a term such as "guess" may occur frequently in all documents, while another term such as "software" may only occur in a few. The reason is that one might make "guesses" in various contexts, regardless of the specific topic, while "software" is a more semantically focused term that is likely to occur only in documents that deal with computer software. A common and very useful transformation that reflects both the semantic specificity of words (document frequencies) as well as the overall frequencies of their occurrences (word frequencies) is the so-called inverse document frequency (for word i and document j):

$$idf\,(i,j) = \begin{cases} 0 & \text{if } wf_{ij}=0 \\ (1+\log(wf_{ij}))\log\frac{N}{df_i} & \text{if } wf_{ij}\geq 1 \end{cases}$$

In this formula, N is the total number of documents, and dfi is the document frequency for the i'th word (the number of documents that include this word). Hence, it can be seen that this formula includes both a dampening of the simple word frequencies via a log function, and also includes a weighting factor that evaluates to 0 if the word occurs in all documents (log(N/N=1)=0), and to the maximum value when a word only occurs in a single document (log(N/1)=log(N)). It can easily be seen how this transformation will create indices that both reflect the relative frequencies-of-occurrences of words, as well as their semantic specificities over the documents included in the analysis.

Irregular component We want to increase our understanding of a time series by picking out its main features. One of these main features is the irregular component (or 'noise'). Descriptive techniques may be extended to forecast (predict) future values.

The irregular component is that left over when the other components of the series (trend, seasonal and cyclical) have been accounted for.

See also time series.

See also trend component.

See also cyclical component.

See also seasonal component.

Ishikawa chart A type of diagram used to depict the factors or variables that make up a process; named after Professor Kaoru Ishikawa of Tokyo University, this diagram is also referred to as the Cause-and-Effect diagram.

ISO [International standards organisation, based in Geneva, Switzerland] This body publishes specifications for a number of standard programming languages. The specifications are arranged generally to concur with those of ANSI.

Isotropic deviation assignment This algorithm uses the isotropic deviation heuristic

(Haykin, 1994) to assign the deviations to radial units. This heuristic attempts to determine a reasonable deviation (the same for all units), based upon the number of centers, and how spread out they are.

This isotropic deviation heuristic sets radial deviations to:

$$\frac{k}{d^2}$$

where d is the distance between the two most distant centers, and k is the number of centers.

IV IV stands for Independent Variable. See also independent vs. dependent variables.

J

Jacobian matrix The first-order derivative of a continuous and differentiable function F (of multiple parameters) is sometimes called the Jacobian matrix J of F (at some specific values of parameter vector x). The Jacobian matrix plays an important role in most computational algorithms for estimating parameter values for nonlinear regression problems, in particular in the Gauss-Newton and Levenberg-Marquardt algorithms.

Jogging weights Adding a small random amount to the weights in a neural network, in an attempt to escape a local optima in error space.

Johnson curves Johnson (1949) described a system of frequency curves that represents transformations of the standard normal curve. By applying these transformations to a standard normal variable, a wide variety of non-normal distributions can be approximated, including distributions which are bounded on either one or both sides (e.g., U- shaped distributions). The advantage of this approach is that once a particular Johnson curve has been fit, the normal integral can be used to compute the expected percentage points under the respective curve.

Join A join shows how data is related between two tables. When two tables contain matching values on a field, records from the two tables can be combined by defining a Join. For example, suppose one table has the weight of objects with their associated part number and another table has part numbers and their associated product names. A join specifies that the two part number fields are equivalent and allows weights and product names to be related.

Joining networks It is sometimes useful to be able to join two networks together to form a single composite network for a number of reasons:

You might train one network to do preprocessing of data, and another to further classify the preprocessed data. Once completed, the networks can be joined together to classify raw data.

You might want to add a loss matrix to a classification network, to make minimum cost decisions.

Note: Networks can only be joined if the number of input neurons in the second network matches the number of output neurons in the first network. The input neurons from the second network are discarded, and their fan-out weights are attached to the output neurons of the first network.

Caution The post-processing information from the first network and the input preprocessing information from the second network are also discarded. The composite network is unlikely to make sense unless you have designed the two networks with this in mind; i.e., with no post-processing performed by the first network and no preprocessing performed by the second network.

Joint occurrence Both outcomes happening simultaneously; P(AB).

Joint probability The probability of the intersection of two (or more) events.

Joint probability distribution If $X_1, X_2, \ldots, X_k$ are random variables, their *joint probability distribution* gives the probability of events determined by the collection of random variables: for any collection of sets of numbers $\{A_1, \ldots, A_k\}$, the joint probability distribution determines

$P((X_1 \text{ is in } A_1) \text{ and } (X_2 \text{ is in } A_2) \text{ and} \ldots \text{and } (X_k \text{ is in } A_k))$.

K

K means algorithm The K-means algorithm (Moody and Darkin, 1989; Bishop, 1995) assigns radial centers to the first hidden layer in the network if it consists of radial units.

K-means assigns each training case to one of K clusters (where K is the number of radial units), such that each cluster is represented by the centroid of its cases, and each case is nearer to the centroid of its cluster than to the centroids of any other cluster. It is the centroids that are copied to the radial units.

The intention is to discover a set of cluster centers which best represent the natural distribution of the training cases.

K nearest algorithm An algorithm to assign deviations to radial units. Each deviation is the mean distance to the K-nearest neighbors of the point.

K nearest neighbor deviation The K-nearest neighbor deviation assignment algorithm (Bishop, 1995) assigns deviations to radial units by using the RMS (Root Mean Squared) distance from the K units closest to (but not coincident with) each unit as the standard deviation (assuming the unit models a Gaussian). Each unit hence has its own independently calculated deviation, based upon the density of points close to itself.

If less than K non-coincident neighbors are available, the algorithm uses the neighbors that are available.

Kaplan Meier estimator (also called product-limit estimator) A procedure for estimating the survivor function from a set of data that includes censored observations. In the absence of of censoring, the Kaplan-Meier estimator just gives one minus the empirical cdf.

Kendall tau Kendall tau is equivalent to the Spearman R statistic with regard to the underlying assumptions. It is also comparable in terms of its statistical power. However, Spearman R and Kendall tau are usually not identical in magnitude because their underlying logic, as well as their computational formulas are very different. Siegel and Castellan (1988) express the relationship of the two measures in terms of the inequality:

-1 < = 3 * Kendall tau - 2 * Spearman R < = 1

More importantly, Kendall tau and Spearman R imply different interpretations: While Spearman R can be thought of as the regular Pearson product-moment correlation coefficient as computed from ranks, Kendall tau rather represents a probability. Specifically, it is the difference between the probability that the observed data are in the same order for the two variables versus the probability that the observed data are in different orders for the two variables. Kendall (1948, 1975), Everitt (1977), and Siegel and Castellan (1988) discuss Kendall tau in greater detail. Two different variants of tau are computed, usually called tau_b and tau_c. These measures differ only with regard as to how tied ranks are handled. In most cases these values will be fairly similar, and when discrepancies occur, it is probably always safest to interpret the lowest value.

Kernel functions Simple functions (typically Gaussians) which are added together, positioned at known data points, to approximate a sampled distribution (Parzen, 1962).

KL nearest neighbor labeling This algorithm assigns labels to units based upon the labels of the K nearest neighboring training cases. Provided that at least L of the K neighbors are of the same class, this class is used to label the unit. If not, a blank label is applied, signifying an "unknown" class.

Note that this is distinct from (although related to) the KL-Nearest algorithm used when executing cluster networks, which reports the class of at least L of the K nearest units to the input case.

Kohonen The Kohonen error assumes that the second layer of the network consists of radial units representing cluster centers. The error is the distance from the input case to the nearest of these. The Kohonen error function is intended for use with Kohonen networks and Cluster networks only.

Kohonen algorithm The Kohonen algorithm (Kohonen, 1982; Patterson, 1996; Fausett, 1994) assigns centers to a radial hidden layer by attempting to recognize clusters within the training cases. Cluster centers close to one another in pattern-space tend to be assigned to units that are close to each other in the network (topologically ordered).

The Kohonen training algorithm is the algorithm of choice for Self Organizing Feature Map networks. It can also be used to train the radial layer in other network types; specifically, radial basis function, cluster, and generalized regression neural networks.

SOFM networks are typically arranged with the radial layer laid out in two dimensions. From an initially random set of centers, the algorithm tests each training case and selects the nearest center. This center and its neighbors are then updated to be more like the training case.

Over the course of the algorithm, the learning rate (which controls the degree of adaptation of the centers to the training cases) and the size of the neighborhood are gradually reduced. In the early phases, therefore, the algorithm assigns a rough topological map, with similar clusters of cases located in certain areas of the radial layer. In later phases the topological map is fine-tuned, with individual units responding to small clusters of similar cases.

If the neighborhood is set to zero throughout, the algorithm is a simple cluster-assignment technique. It can also be used on a one-dimensional layer with or without neighborhood definition.

If class labels are available for the training cases, then after Kohonen training, labels can be assigned using class labeling algorithms and Learned Vector Quantization used to improve the positions of the radial exemplars.

Kohonen networks Neural networks based on the topological properties of the human brain, also known as self-organizing feature maps (SOFMs).

Kohonen training An algorithm which assigns cluster centers to a radial layer by iteratively submitting training patterns to the network, and adjusting the winning (nearest) radial unit center, and its neighbors, towards the training pattern.

Kolmogorov Smirnov test The Kolmogorov-Smirnov one-sample test for normality is based on the maximum difference between the sample cumulative distribution and the hypothesized cumulative distribution. If the D statistic is significant, then the hypothesis that the respective distribution is normal should be rejected. For many software programs, the probability values that are reported are based on those tabulated by Massey (1951); those probability values are valid when the mean and standard deviation of the normal distribution are known a-priori and not estimated from the data. However, usually those parameters are computed from the actual data. In that case, the test for normality involves a complex conditional hypothesis ("how likely is it to obtain a D statistic of this magnitude or greater, contingent upon the mean and standard deviation computed from the data"), and the Lilliefors probabilities should be interpreted (Lilliefors, 1967). Note that in recent years, the Shapiro-Wilks' W test has become the preferred test of normality because of its good power properties as compared to a wide range of alternative tests.

Kronecker product The Kronecker (direct) product of 2 matrices A, with p rows and q columns, and B, with m rows and n columns, is the matrix with pm rows and qn columns given by

A ? B = a_{ij} B

For example, if

$$\mathbf{A} = \begin{bmatrix} 1 & 0 \\ 0 & 1 \\ -1 & -1 \end{bmatrix}$$

and

$$\mathbf{B} = \begin{bmatrix} 1 \\ -1 \end{bmatrix}$$

then

$$\mathbf{A} \otimes \mathbf{B} = \begin{bmatrix} 1 & 0 \\ 0 & 1 \\ -1 & -1 \\ -1 & 0 \\ 0 & -1 \\ 1 & 1 \end{bmatrix}$$

Kronecker Product matrices have a number of useful properties.

Kruskal Wallis test The Kruskal-Wallis test is a non-parametric test used to compare three or more samples. It is used to test the null hypothesis that all populations have identical distribution functions against the alternative hypothesis that at least two of the samples differ only with respect to location (median), if at all.

It is the analogue to the F-test used in analysis of variance. While analysis of variance tests depend on the assumption that all populations under comparison are normally distributed, the Kruskal-Wallis test places no such restriction on the comparison.

It is a logical extension of the Wilcoxon-Mann-Whitney Test.

Kurtosis A property of a distribution which reflects the 'peakedness' of the plotted curve. A tall, peaked distribution is called leptokurtic while a flat, plateau-like shaped distribution is called platykurtic. A symmetrical curve is called mesokurtic. Kurtosis (the term first used by Pearson, 1905) measures the "peakedness" of a distribution. If the kurtosis is clearly different than 0, then the distribution is either flatter or more peaked than normal.

L

Lack of fit For certain designs with replicates at the levels of the predictor variables, the residual sum of squares can be further partitioned into meaningful parts which are relevant for testing hypotheses. Specifically, the residual sums of squares can be partitioned into lack-of-fit and pure-error components. This involves determining the part of the residual sum of squares that can be predicted by including additional terms for the predictor variables in the model (for example, higher-order polynomial or interaction terms), and the part of the residual sum of squares that cannot be predicted by any additional terms (i.e., the sum of squares for pure error). A test of lack-of-fit for the model without the additional terms can then be performed, using the mean square pure error as the error term. This provides a more sensitive test of model fit, because the effects of the additional higher-order terms is removed from the error.

Lambda prime Lambda is defined as the geometric sum of 1 minus the squared canonical correlation, where lambda is Wilk's lambda. The squared canonical correlation is an estimate of the common variance between two canonical variates, thus 1 minus this value is an estimate of the unexplained variance. Lambda is used as a test of significance for the squared canonical correlation and is distributed as Chi-square.

$$\chi^2 = [-N -1 - \{.5(p+q+1)\}] * \log_e \lambda$$

where N is the number of subjects

p is the number of variables on the right

q is the number of variables on the left

Laplace distribution The Laplace (or double exponential) distribution has density function:

$$f(x) = 1/(2b)*e^{-|x-a|/b} \quad -\infty < x < \infty$$

where a is the mean of the distribution

b is the scale parameter

e is the base of the natural logarithm, sometimes called Euler's E.

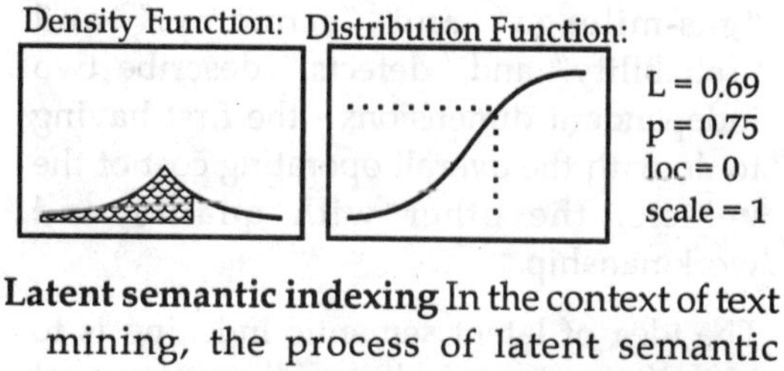

Latent semantic indexing In the context of text mining, the process of latent semantic indexing is concerned with the derivation of underlying dimensions of "meaning" from the words (terms) extracted from a collection of documents.

The most basic result of text mining is an initial indexing of words found in the input documents, and the computation of a frequency table with simple counts enumerating the number of times that each word occurs in each input document. Also, in practice, you can further transform those raw counts to indices that better reflect the (relative) "importance" of words and/or their semantic specificity in the context of the set of input documents.

Next, a common analytic tool for interpreting the "meaning" or "semantic space" described by the words that were extracted and, hence, by the documents that were analyzed, is to create a mapping of the words and documents into a common space, computed from the word frequencies or transformed word frequencies (e.g., inverse document frequencies). In general, here is how it works:

Suppose you index a collection of customer reviews of their new automobiles (e.g., for different makes and models). You may find that every time a review includes the word "gas-mileage," it also includes the term "economy." Further, when reports include the word "reliability" they also include the term "defects" (e.g., make reference to "no defects"). However, there is no consistent pattern regarding the use of the terms "economy" and "reliability," i.e., some documents include either one, both, or neither. In other words, these four words "gas-mileage" and "economy," and "reliability" and "defects," describe two independent dimensions - the first having to do with the overall operating cost of the vehicle, the other with quality and workmanship.

The idea of latent semantic indexing is to identify such underlying dimensions (of "meaning"), into which the words and documents can be mapped. As a result, you can identify the underlying (latent) themes described or discussed in the input documents, and also identify the documents that mostly deal with each dimension (e.g., economy, reliability, or both). In practice, singular value decomposition is often used to extract the underlying semantic dimensions from the matrix of (transformed) word counts across documents.

Latent variable A latent variable is a variable that cannot be measured directly, but is hypothesized to underlie the observed variables. An example of a latent variable is a factor in factor analysis. Latent variables in path diagrams are usually represented by a variable name enclosed in an oval or circle.

Law of averages The Law of Averages says that the average of independent observations of random variables that have the same probability distribution is increasingly likely to be close to the expected value of the random variables as the number of observations grows. More precisely, if X_1, X_2, X_3, . . ., are independent random variables with the same probability distribution, and E(X) is their common expected value, then for every number E > 0,

$$P\{ | (X_1 + X_2 + \ldots + X_n)/n - E(X) | < E\}$$

converges to 100% as n grows. This is equivalent to saying that the sequence of sample means

$$X_1, (X_1+X_2)/2, (X_1+X_2+X_3)/3,$$

converges in probability to E(X).

Law of large numbers The Law of Large Numbers says that in repeated, independent trials with the same probability p of success in each trial, the percentage of successes is increasingly likely to be close to the chance of success as the number of trials increases. More precisely, the chance that the percentage of successes differs from the probability p by more than a fixed positive amount, E > 0, converges to zero as the number of trials n goes to infinity, for every number $e > 0$. Note that in contrast to the difference between the *percentage* of successes and the probability of success, the difference between the *number* of successes and the expected number of successes, $n \times p$, tends to grow as n grows.

Layered compression When layered compression is used, the main graph plotting area is reduced in size to leave space for Margin Graphs in the upper and right side of the display (and a miniature graph in the corner). These smaller Margin Graphs represent vertically and horizontally compressed images (respectively) of the main graph.

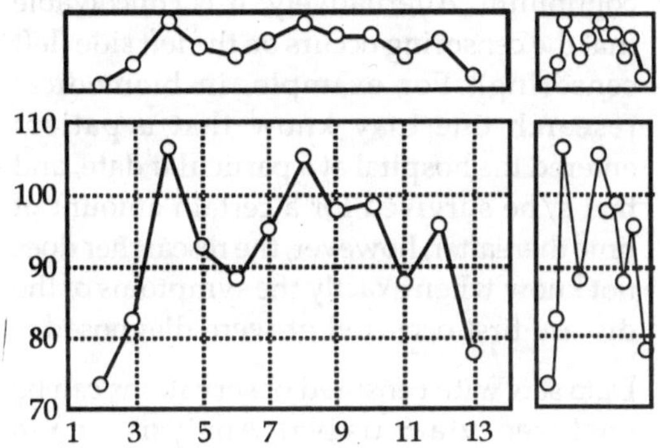

Learned vector quantization The learned vector quantization algorithm (LVQ) was invented by Tuevo Kohonen (Fausett, 1994; Kohonen, 1990), who also invented the Self-Organizing Feature Map.

Learned Vector Quantization provides a supervised version of the Kohonen training algorithm. The standard Kohonen algorithm iteratively adjusts the position of the exemplar vectors stored in the Radial layer of the Kohonen network by considering only the positions of the existing vectors and of the training data. In essence, the algorithm attempts to move the exemplar vectors to positions that reflect the centers of clusters in the training data. However, the class labels of the training data cases are not taken into account. For superior classification performance, it is desirable that the exemplar vectors are adjusted, to some extent, on a per-class basis - that is, that they reflect natural clusters in each separate class. An exemplar located on a class boundary, equally close to cases of two classes, is unlikely to be of much use in distinguishing class. On the other hand, exemplars located just inside class boundaries can be extremely useful.

There are several variants of Learned Vector Quantization. The basic version, LVQ1, is very similar to the Kohonen training algorithm. The closest exemplar to a training case is selected during training and has its position updated. However, whereas the Kohonen algorithm would move this exemplar toward the training case, LVQ1 checks whether the class label of the exemplar vector is the same as that of the training case. If it is, the exemplar is moved toward the training case; if it is not, the exemplar is moved away from the case. The more sophisticated LVQ algorithms, LVQ2.1 and LVQ3, take into account more information. They locate the nearest two exemplars to the training case. If one of these is of the right class and one the wrong class, they move the right class toward the training case and the wrong one away from it. LVQ3 also moves both exemplars toward the training case if they are both of the right class. In both LVQ2.1 and LVQ3, the concept is to move exemplars where there is some danger of misclassification.

Learning rate A control parameter of some training algorithms, which controls the step size when weights are iteratively adjusted.

Least squares The method of least squares is a criterion for fitting a specified model to observed data. For example, it is the most commonly used method of defining a straight line through a set of points on a scatterplot.

Least squares (2D graphs) A curve is fitted to the XY coordinate data according to the distance-weighted least squares smoothing procedure (the influence of individual points decreases with the horizontal distance from the respective points on the curve).

Least squares (3D graphs) A surface is fitted to the XYZ coordinate data according to the distance-weighted least squares smoothing procedure (the influence of individual points decreases with the horizontal distance from the respective points on the surface).

Least squares estimator In the most general terms, least squares estimation is aimed at minimizing the sum of squared deviations of the observed values for the dependent variable from those predicted by the model. Technically, the least squares estimator of a parameter q is obtained by minimizing Q with respect to q where:

$$Q = \Sigma [Y_i - f_i(\theta)]^2$$

Note that $f_i(\theta)$ is a known function of θ, $Y_i = f_i(\theta) + \varepsilon_i$ where $i = 1$ to n, and the ε_i are random variables, and usually assumed to have expectation of 0.

Least squares line The line passing through the points that minimize the sum of squared deviations: A synonym for the regression line.

Least squares means When there are no missing cells in ANOVA designs with categorical predictor variables, the subpopulation (or marginal) means are least square means, which are the best linear-unbiased estimates of the marginal means for the design. Tests of differences in least square means have the important property that they are invariant to the choice of the coding of effects for categorical predictor variables (e.g., the use of the sigma-restricted or the overparameterized model) and to the choice of the particular generalized inverse of the design matrix used to solve the normal equations. Thus, tests of linear combinations of least square means in general are said to not depend on the parameterization of the design.

Left and right censoring When censoring, a distinction can be made to reflect the "side" of the time dimension at which censoring occurs. Consider an experiment where we start with 100 light bulbs, and terminate the experiment after a certain amount of time. In this experiment the censoring always occurs on the right side (right censoring), because the researcher knows when exactly the experiment started, and the censoring always occurs on the right side of the time continuum. Alternatively, it is conceivable that the censoring occurs on the left side (left censoring). For example, in biomedical research one may know that a patient entered the hospital at a particular date, and that s/he survived for a certain amount of time thereafter; however, the researcher does not know when exactly the symptoms of the disease first occurred or were diagnosed.

Data sets with censored observations can be analyzed via Survival Analysis or via Weibull and Reliability/Failure Time Analysis.

Level of measurement The name given to the classification scheme which distinguishes the relationships between the categories or attributes of a variable. See nominal, ordinal, interval and ratio variables.

Level of significance The probability that the research outcome could have happened by chance

Levenberg Marquardt algorithm Levenberg-Marquardt (Levenberg, 1944; Marquardt, 1963; Bishop, 1995; Shepherd, 1997; Press et. al., 1992) is an advanced non-linear optimization algorithm. It can be used to train the weights in a network just as back propagation would be. It is reputably the fastest algorithm available for such training.

Levene and Brown Forsythe tests for homogeneity of variances (HOV) A important assumption in analysis of variance (ANOVA and the t-test for mean differences) is that the variances in the different groups are equal (homogeneous). Two powerful and commonly used tests of this assumption are the Levene test and the Brown-Forsythe modification of this test. However, it is important to realize that 1. The homogeneity of variances assumption is usually not as crucial as other assumptions for ANOVA, in particular in the case of balanced (equal n), and

2. That the tests described are not necessarily very robust themselves.

Levene's test (homogeneity of variances) For each dependent variable, an analysis of variance is performed on the absolute deviations of values from the respective group means. If the Levene test is statistically significant, then the hypothesis of homogeneous variances should be rejected.

Leverage A measure of the extent to which an outlier is both distant from the regression line and has a large value on an independent variable.

Leverage values In regression, this term refers to the diagonal elements of the hat matrix $(X(X'X)^{-1}X')$. A given diagonal element $(h_{(ii)})$ represents the distance between X values for the ith observation and the means of all X values. These values indicate whether or not X values for a given observation are outlying. The diagonal element is refered to as the leverage. A large leverage value indicates that the ith observation is distant from the center of the X observations (Neter, et al, 1985).

Life table The most straightforward way to describe the survival in a sample is to compute the Life Table. The life table technique is one of the oldest methods for analyzing survival (failure time) data. This table can be thought of as an "enhanced" frequency distribution table. The distribution of survival times is divided into a certain number of intervals. For each interval one can compute the number and proportion of cases or objects that entered the respective interval "alive," the number and proportion of cases that failed in the respective interval (i.e., number of terminal events, or number of cases that "died"), and the number of cases that were lost or censored in the respective interval.

Based on those numbers and proportions, several additional statistics can be computed.

Lift chart The lift chart provides a visual summary of the usefulness of the information provided by one or more statistical models for predicting a binomial (categorical) outcome variable (dependent variable); for multinomial (multiple-category) outcome variables, lift charts can be computed for each category. Specifically, the chart summarizes the utility that one may expect by using the respective predictive models, as compared to using baseline information only.

The lift chart is applicable to most statistical methods that compute predictions (predicted classifications) for binomial or multinomial responses. This and similar summary charts are commonly used in Data Mining projects, when the dependent or outcome variable of interest is binomial or multinomial in nature.

Example. To illustrate how the lift chart is constructed, consider this example. Suppose you have a mailing list of previous customers of your business, and you want to offer to those customers an additional service by mailing an elaborate brochure and other materials describing the service. During previous similar mail-out campaigns, you collected useful information about your customers (e.g., demographic information, previous purchasing patterns) that you could relate to the response rate, i.e., whether the respective customers responded to your mail solicitation. Also, from similar prior mail-out campaigns, you were able to estimate the baseline response rate at approximately 7 percent, i.e., 7% of all customers who received a similar offer by mail responded (purchased the additional service).

Given this baseline response rate (7%) and the cost of the mail-out, sending the offer to all customers would result in a net loss. Hence, you want to use statistical analyses to help you identify the customers who are most likely to respond. Suppose you build such a model based on the data collected in the previous mail-out campaign. You can now select only the 10 percent of the customers from the mailing lists who, according to prediction from the model, are most likely to respond. If among those customers (selected by the model) the response rate is 14% percent (as opposed to the 7% baseline rate), then the relative gain or lift value due to using the predictive model can be computed as 14% / 7% = 2. In other words, you were able to do twice as well as you would have done using simple random selection.

Analogous lift values can be computed for each percentile of the population (customers on the mailing list). You could compute separate lift values for selecting the top 20% of customers who are predicted to be among likely responders to the mail campaign, the top 30%, etc. Hence, the lift values for different percentiles can be connected by a line that will typically descend slowly and merge with the baseline if all customers (100%) would be selected.

If more than one predictive model is used, multiple lift charts can be overlaid to provide a graphical summary of the utility of different models.

Likelihood The probability of an event based on current observations.

Likelihood function (often called just likelihood.) A function expressing the value of a pmf (or probability density) at an observed value of a random variable in terms of the values of one or more unknown parameters.

Lilliefors test In a Kolmogorov-Smirnov test for normality when the mean and standard deviation of the hypothesized normal distribution are not known (i.e., they are estimated from the sample data), the probability values tabulated by Massey (1951) are not valid. Instead, the so-called Lilliefors probabilities (Lilliefors, 1967) should be used in determining whether the KS difference statistic is significant.

Limit See *converge.*

Line graph A graph used to plot data showing the relationship between independent and dependent variables in an experiment.

Line plots, 2D (case profiles) Unlike in the regular line plots where the values of one variable are plotted as one line (individual data points are connected by a line), in case profile line plots, the values of the selected variables in a case (row) are plotted as one line (i.e., one line plot will be generated for each of the selected cases). Case profile line plots provide a simple way to visually present the values in a case (e.g., test scores for several tests).

Line plots, 2D aggregrated Aggregated line plots display a sequence of means for consecutive subsets of a selected variable.

You can select the number of consecutive observations from which the mean will be calculated and if desired, the range of values in each subset will be marked by the whisker-type markers . Aggregated line plots are used to explore and present sequences of large numbers of values.

Line plots, 2D double Y The Double-Y line plot can be considered to be a combination of two separately scaled multiple line plots. A separate line pattern is plotted for each of the selected variables, but the variables selected in the Left-Y list will be plotted against the left-Y axis, whereas the variables selected in the Right-Y list will be plotted against the right-Y axis. The names of all variables will be identified in the legend with the letter (R) for the variables associated with the right-Y axis and with the letter (L) for the variables associated with the left-Y axis.

The Double-Y line plot can be used to compare sequences of values of several variables by overlaying their respective line representations in a single graph. However, due to the independent scaling used for the two axes, it can facilitate comparisons between otherwise "incomparable" variables (i.e., variables with values in different ranges).

Line plots, 2D In line plots, individual data points are connected by a line.

Line plots provide a simple way to visually present a sequence of values. XY Trace-type line plots can be used to display a trace (instead of a sequence). Line plots can also be used to plot continuous functions, theoretical distributions, etc.

Line plots, 2D multiple Unlike regular line plots in which a sequence of values of one variable is represented, the multiple line plot represents multiple sequences of values (variables). A different line pattern and color is used for each of the multiple variables and referenced in the legend.

This type of line plot is used to compare sequences of values between several variables (or several functions) by overlaying them in a single graph that uses one common set of scales (e.g., comparisons between several simultaneous experimental processes, social phenomena, stock or commodity quotes, shapes of operating characteristics curves, etc.).

Line plots, 2D regular Regular line plots are used to examine and present the sequences of values (usually when the order of the presented values is meaningful).

Another typical application for line sequence plots is to plot continuous functions, such as fitted functions or theoretical distributions. Note that an empty data cell (i.e., missing data) "breaks" the line.

Line plots, 2D XY trace In trace plots, a scatterplot of two variables is first created, then the individual data points are connected with a line (in the order in which they are read from the data file). In this sense, trace plots visualize a "trace" of a sequential process (movement, change of a phenomenon over time, etc.)

Linear (2D graphs) A linear function (e.g., Y = a + bX) is fitted to the points in the 2D scatterplot.

Linear (3D graphs) A linear function (e.g., Y = a + bX) is fitted to the points in the 3D scatterplot.

Linear activation function A null activation function: the unit's output is identical to its activation level.

Linear association Two variables are linearly associated if a change in one is associated with a proportional change in the other, with the same constant of proportionality throughout the range of measurement. The correlation coefficient measures the degree of linear association on a scale of -1 to 1.

Linear modelling Approximation of a discriminant function or regression function using a hyperplane. Can be globally optimized using "simple" techniques, but does not adequately model many real-world problems.

Linear operation Suppose *f* is a function or operation that acts on things we shall denote generically by the lower-case Roman letters *x* and *y*. Suppose it makes sense to multiply *x* and *y* by numbers (which we denote by *a*), and that it makes sense to add things like *x* and *y* together. We say that *f* is *linear* if for every number *a* and every value of *x* and *y* for which $f(x)$ and $f(y)$ are defined, (i) $f(a \times x)$ is defined and equals $a \times f(x)$, and (ii) $f(x + y)$ is defined and equals $f(x) + f(y)$. *C.f.* affine.

Linear units A unit with a linear PSP function. The unit's activation level is the weighted sum of its inputs minus the threshold - also known as a dot product or linear combination. The characteristic unit type of multilayer perceptrons. Despite the name, a linear unit may have a non-linear activation function.

Local minima Local "valleys" or minor "dents" in a loss function which, in many practical applications, will produce extremely large or small parameter estimates with very large standard errors. The Simplex method is particularly effective in avoiding such minima; therefore, this method may be particularly well suited in order to find appropriate start values for complex functions.

Location, measure of A measure of location is a way of summarizing what a "typical" element of a list is—it is a one-number summary of a distribution. See also arithmetic mean, median, and mode.

Log linear analysis Log-linear analysis provides a "sophisticated" way of looking at crosstabulation tables (to explore the data or verify specific hypotheses), and it is sometimes considered an equivalent of ANOVA for frequency data. Specifically, it allows the user to test the different factors that are used in the crosstabulation (e.g., gender, region, etc.) and their interactions for statistical significance.

Log normal distribution The lognormal distribution (the term first used by Gaddum, 1945) has the probability density function:

$$f(x) = 1/[x\sigma (2\pi)^{1/2}] * \exp(-[\log(x)-\mu]^2/2\sigma^2)$$

$0 \le x < \infty$

$\mu > 0$

$\sigma > 0$

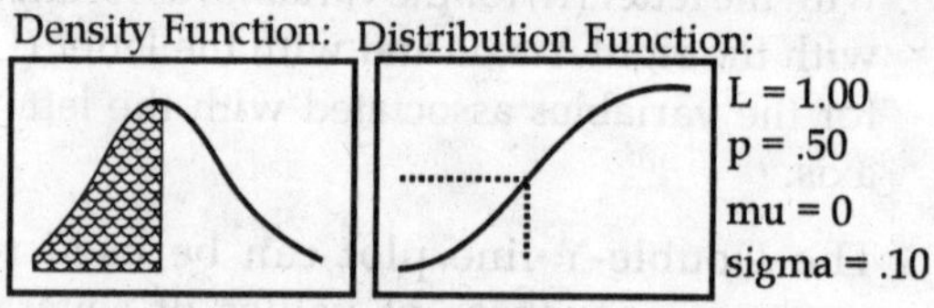

Where μ is the scale parameter

σ is the shape parameter

e is the base of the natural logarithm, sometimes called Euler's e

It shows the Log-normal distribution with mu equal to 0 for sigma equals .10, .30, .50, .70, and .90.

Log rank test A hypothesis test for the null hypothesis that two survivor function are identical against the alternative that the hazard of one is a constant multiple of the hazard of the other, the so-called proportional hazards model.

Logarithmic function This fits to the data, a logarithmic function of the form:

$$y = q*[\log_n(x)] + b$$

Logistic This is an S-shaped (sigmoid) curve, with output in the range (0,1).

Logistic distribution The Logistic distribution has density function:

$f(x) = (1/b)*e^{-(x-a)/b} * [1+e^{-(x-a)/b}]^{-2}$

where a is the mean of the distribution

b is the scale parameter e is the base of the natural logarithm, sometimes called Euler's E.

Logistic function Also know as logistic distribution, or logistic cdf.

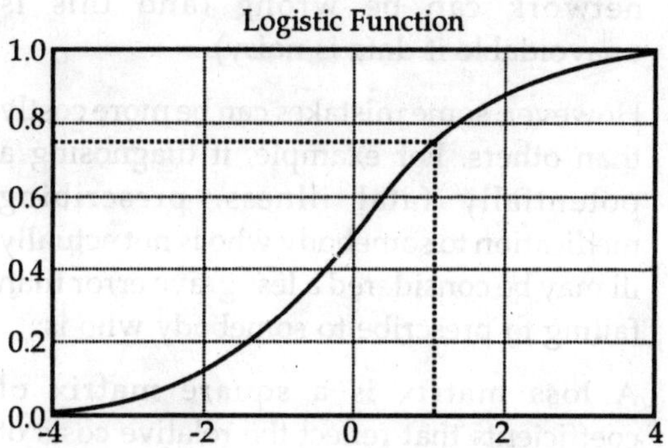

The inverse of the logit transform, giving the probability corresponding to a given logit.

Logistic regression This relates to an experimental design for predicting a binary categorical (yes/no) outome on the basis of predictor variables measured on interval scales. For each of a set of values of the predictor variables, the outcomes are regarded as representing a binomial process, with the binomial parameter 'p' depending upon the value of the predictor variable. The modelling accounts for the logarithm of the odds ratio as a linear function of the predictor variable. Fitting is via a weighted least-squares regression method. Randomisation tests for this purpose have been developed by Mehta & Patel.

Logit Also called log-odds The (natural) logarithm of the odds in favour of an event. The logit of a binary variable is the log-odds that its value is a "success".

Logit regression and transformation In the logit regression model, the predicted values for the dependent or response variable will never be less than (or equal to) 0, or greater than (or equal to) 1, regardless of the values of the independent variables; it is, therefore, commonly used to analyze binary dependent or response variables (see also the binomial distribution). This is accomplished by applying the following regression equation (the term logit was first used by Berkson, 1944):

$y=exp(b_0 +b_1*x_1 + ... + b_n*x_n)/\{1+exp(b_0 +b_1*x_1 + ... + b_n*x_n)\}$

One can easily recognize that, regardless of the regression coefficients or the magnitude of the x values, this model will always produce predicted values (predicted y's) in the range of 0 to 1. The name logit stems from the fact that one can easily linearize this model via the logit transformation. Suppose we think of the binary dependent variable y in terms of an underlying continuous probability p, ranging from 0 to 1. We can then transform that probability p as:

$p' = \log_e\{p/(1-p)\}$

This transformation is referred to as the logit or logistic transformation. Note that p' can theoretically assume any value between minus and plus infinity. Since the logit transform solves the issue of the 0/1 boundaries for the original dependent variable (probability), we could use those (logit transformed) values in an ordinary linear regression equation. In fact, if we perform the logit transform on both sides of the logit regression equation stated earlier, we obtain the standard linear multiple regression model:

$p' = (b_0 +b_1*x_1 +...+ b_n*x_n)$

See also Probit Transformation and Regression and Multinomial logit and probit regression for similar transformations.

Longitudinal data Data collected from the same people over a period of time.

Longitudinal study A study in which individuals are followed over time, and compared with themselves at different times, to determine, for example, the effect of aging on some measured variable. Longitudinal studies provide much more persuasive evidence about the effect of aging than do cross-sectional studies.

Lookahead For neural networks time series analysis, the number of time steps ahead of the last input variable values the output variable values should be predicted.

Loss function The loss function (the term loss was first used by Wald, 1939) is the function that is minimized in the process of fitting a model, and it represents a selected measure of the discrepancy between the observed data and data "predicted" by the fitted function. For example, in many traditional general linear model techniques, the loss function (commonly known as least squares) is the sum of squared deviations from the fitted line or plane. One of the properties (sometimes considered to be a disadvantage) of that common loss function is that it is very sensitive to outliers.

A common alternative to the least squares loss function is to maximize the likelihood or log-likelihood function (or to minimize the negative log-likelihood function; the term maximum likelihood was first used by Fisher, 1922a). These functions are typically used when fitting non-linear models. In most general terms, the likelihood function is defined as:

$$L=F(Y,Model)=\prod_{i=1}^{n} \{ p[y_i , Model Parameters(x_i)]\}$$

In theory, we can compute the probability (now called L, the likelihood) of the specific dependent variable values to occur in our sample, given the respective regression model.

Loss matrix If a network is trained so that the outputs estimate probabilities, it can be adjusted to support a loss matrix (Bishop, 1995).

In simple cases, a probability estimate may be used directly: an unknown case is simply assigned to the most-probable class. Inevitably, this means that sometimes the network can be wrong (and this is unavoidable if data is noisy).

However, some mistakes can be more costly than others. For example, if diagnosing a potentially fatal illness, prescribing medication to somebody who is not actually ill may be considered a less grave error than failing to prescribe to somebody who is.

A loss matrix is a square matrix of coefficients that reflect the relative costs of various misclassifications. It is multiplied by the vector of probability estimates, resulting in a vector of cost estimates, and the case is assigned to the class with the lowest cost estimate.

Since a correct classification has zero cost, the leading diagonal of a loss matrix always contains zeros; in other positions, the coefficient in the n'th column and m'th row represents the cost of misclassifying a case that is actually in the n'th class as being in the m'th class.

Lower confidence limit The lower endpoint of a confidence interval. The term is typically used when the upper enpdoint is infinity.

Lower fence A value equal to the lower quartile less 1.5×IQR.

Lower quartile (Q1) A value that exceeds the values for no more than one quarter of the individuals and is exceeded by the values of no more than three quarters of the individuals. cf. upper quartile.) The 25th percentile of a set of measures.

Lowess smoothing (robust locally weighted regression) Robust locally weighted regression is a method of smoothing 2D scatterplot data (pairs of x-y data). A local polynomial regression model is fit to each point and the points close to it. The method is also sometimes referred to as lowess smoothing. The smoothed data usually provide a clearer picture of the overall shape of the relationship between the x and y variables.

M

M

Machine learning Machine learning, computational learning theory, and similar terms are often used in the context of Data Mining, to denote the application of generic model-fitting or classification algorithms for predictive data mining. Unlike traditional statistical data analysis, which is usually concerned with the estimation of population parameters by statistical inference, the emphasis in data mining (and machine learning) is usually on the accuracy of prediction (predicted classification), regardless of whether or not the "models" or techniques that are used to generate the prediction is interpretable or open to simple explanation. A good example of this type of technique often applied to predictive data mining are neural networks or meta-learning techniques such as boosting, etc. These methods usually involve the fitting of very complex "generic" models, that are not related to any reasoning or theoretical understanding of underlying causal processes; instead, these techniques can be shown to generate accurate predictions or classification in crossvalidation samples.

Mahalanobis distance One can think of the independent variables (in a regression equation) as defining a multidimensional space in which each observation can be plotted. Also, one can plot a point representing the means for all independent variables. This "mean point" in the multidimensional space is also called the centroid. The Mahalanobis distance is the distance of a case from the centroid in the multidimensional space, defined by the correlated independent variables (if the independent variables are uncorrelated, it is the same as the simple Euclidean distance). Thus, this measure provides an indication of whether or not an observation is an outlier with respect to the independent variable values.

See also, standard residual value, deleted residual and Cook's distance.

Mallow's CP If p regressors are selected from a set of k, Cp is defined as:

$(y-yp)^2 / s^2 - n+2p$

where y_p is the predicted value of y from the p regressors

s^2 is the residual mean square after regression on the complete set of k n is the sample size

The model is then chosen to give a minimum value of the criterion, or a value that is acceptably small. It is essential a special case of Akaike Information Criterion. Mallow's

CP is used in General Regression Models (GRM) as the criterion for choosing the best subset of predictor effects when a best subset regression analysis is being performed. This measure of the quality of fit for a model tends to be less dependent (than the R-square) on the number of effects in the model, and hence, it tends to find the best subset that includes only the important predictors of the respective dependent variable.

Manifest variable A manifest variable is a variable that is directly observable or measurable. In path analysis diagrams used in structural modeling manifest variables are usually represented by enclosing the variable name within a square or a rectangle.

Mann Scheuer Fertig test This test, proposed by Mann, Scheuer, and Fertig (1973), is described in detail in, for example, Dodson (1994) or Lawless (1982). The null hypothesis for this test is that the population follows the Weibull distribution with the estimated parameters. Nelson (1982) reports this test to have reasonably good power, and this test can be applied to Type II censored data. The critical values for the test statistic have been computed based on Monte Carlo studies, and have been tabulated for n (sample sizes) between 3 and 25; for n greater than 25, this test is not computed.

Mann Whitney test This is a test of difference in location for an experimental design involving two samples with data measured on an ordinal scale or better. The test statistic is a measure of ordinal precedence. For each possible pairing of an observation in one group with an observation in the alternate group, the pair is classified in one of three ways - according to whether the difference is positive, zero or negative; the numbers in these three categories are tallied over the randomisation set. The randomisation set is the same as that for the pitman permutation test 1.. This test is generally recommended for comparisons involving ordinal-scale data but is not confined to this scale-type. An equivalent formulation of the test, based upon ranking the data and summing ranks within groups, is the Wilcoxon test

2. Also see combinations.

Mantel Haenszel statistic A test statistic having an approximate chi-squared distribution (with one degree of freedom) used in testing independence in stratified 2×2 tables.

Margin Any one of the row totals, column totals, or grand total.

Margin of error A measure of the uncertainty in an estimate of a parameter; unfortunately, not everyone agrees what it should mean. The *margin of error* of an estimate is typically one or two times the estimated standard error of the estimate.

Marginal The sum of the counts for a particular category of a variable. These sums are often placed in the right-most column or bottom row of a table and are therefore in the margins.

Marginal frequencies In a Multi-way table, the values in the margins of the table are simply one-way (frequency) tables for all values in the table. They are important in that they help us to evaluate the arrangement of frequencies in individual columns or rows. The differences between the distributions of frequencies in individual rows (or columns) and in the respective margins inform us about the relationship between the crosstabulated variables.

Marginal probability The probability of an event when considered as being obtained by adding up joint probabilities. The term marginal probability is used in contexts where it is necessary to distinguish it from joint and/or conditional probabilities.

Markov Chain Monte Carlo (MCMC) The term "Monte Carlo method" (suggested by John von Neumann and S. M. Ulam, in the

1940s) refers to simulation of processes, using random numbers. The term Monte Carlo (a city long known for its gambling casinos) derived from the fact that "numbers of chance" (i.e., Monte Carlo simulation methods) were used in order to solve some of the integrals of the complex equations involved in the design of the first nuclear bombs (integrals of quantum dynamics). By generating large samples of random numbers from, for example, mixtures of distributions, the integrals of these (complex) distributions can be approximated from the (generated) data.

Complex equations with difficult to solve integrals are often involved in Bayesian Statistics Analyses.

Markov's inequality For lists: If a list contains no negative numbers, the fraction of numbers in the list at least as large as any given constant $a>0$ is no larger than the arithmetic mean of the list, divided by a. For random variables: if a random variable X must be nonnegative, the chance that X exceeds any given constant $a>0$ is no larger than the expected value of X, divided by a.

Mass The term mass in correspondence analysis is used to denote the entries in the two-way table of relative frequencies (i.e., each entry is divided by the sum of all entries in the table). Note that the results from correspondence analysis are still valid if the entries in the table are not frequencies, but some other measure of correspondence, association, similarity, confusion, etc. Since the sum of all entries in the table of relative frequencies is equal to 1.0, one could say that the table of relative frequencies shows how one unit of mass is distributed across the cells of the table. In the terminology of correspondence analysis, the row and column totals of the table of relative frequencies are called the row mass and column mass, respectively.

Matched pair design A study design in which each individual is paired with another individual that resembles it in all respects except the property being studied.

Matched samples a. Two samples in which the members are clearly paired, or are matched explicitly by the researcher. For example, IQ measurements on pairs of identical twins.

b. Those samples in which the same attribute, or variable, is measured twice on each subject, under different circumstances. Commonly called repeated measures. Examples include the times of a group of athletes for 1500m before and after a week of special training; the milk yields of cows before and after being fed a particular diet.

Sometimes, the difference in the value of the measurement of interest for each matched pair is calculated, for example, the difference between before and after measurements, and these figures then form a single sample for an appropriate statistical analysis.

Matching moments method This method can be employed to determine parameter estimates for a distribution (see Quantile-Quantile Plots, Probability-Probability Plots, and Process Analysis). The method of matching moments sets the distribution moments equal to the data moments and solves to obtain estimates for the distribution parameters. For example, for a distribution with two parameters, the first two moments of the distribution (the mean and variance of the distribution, respectively, e.g., μ and σ, respectively) would be set equal to the first two moments of the data (the sample mean and variance, respectively, e.g., the unbiased estimators x-bar and s**2, respectively) and solved for the parameter estimates. Alternatively, you could use the Maximum Likelihood Method to estimate the parameters.

Matrix collinearity, multicollinearity This term is used in the context of correlation matrices or covariance matrices, to describe the condition when one or more variables from which the respective matrix was computed are linear functions of other variables; as a consequence such matrices cannot be inverted (only the generalized Inverse can be computed).

Matrix ill conditioning Matrix ill-conditioning is a general term used to describe a rectangular matrix of values which is unsuitable for use in a particular analysis.

This occurs perhaps most frequently in applications of linear multiple regression when the matrix of correlations for the predictors is singular and thus the regular matrix inverse cannot be computed. In some modules (i.e., in Factor Analysis) this problem is dealt with by issuing a respective warning and then artificially lowering all correlations in the correlation matrix by adding a small constant to the diagonal elements of the matrix, and then restandardizing it. This procedure will usually yield a matrix for which the regular matrix inverse can be computed.

Note that in many applications of the general linear model and the generalized linear model, matrix singularity is not abnormal (i.e., when the overparameterized model is used to represent effects for categorical predictor variables) and is dealt with by computing a generalized inverse rather than the regular matrix inverse.

Another example of matrix ill-conditioning is intransitivity of the correlations in a correlation matrix. If in a correlation matrix variable A is highly positively correlated with B, B is highly positively correlated with C, and A is highly negatively correlated with C, this "impossible" pattern of correlations signals an error in the elements of the matrix.

Matrix inverse The regular inverse of a rectangular matrix of values is an extension of the concept of a numeric reciprocal. For a nonsingular matrix A, its inverse (denoted by a superscript of -1) is the unique matrix that satisfies

$A^{-1}AA=A$

No such regular inverse exists for singular matrices, but generalized inverses (an infinite number of them) can be computed for any singular matrix.

Matrix plots Matrix graphs summarize the relationships between several variables in a matrix of true X-Y plots. The most common type of matrix plot is the scatterplot, which can be considered to be the graphical equivalent of the correlation matrix.

Matrix plots columns In this type of Matrix Plot, columns represent projections of individual data points onto the X- axis (showing the distribution of the maximum values), arranged in a matrix format. Histograms representing the distribution of each variable are displayed along the diagonal of the matrix or along the edges (in rectangular matrices).

Matrix plots lines In this type of Matrix Plot, a matrix of X-Y (i.e., nonsequential) line plots (similar to a scatterplot matrix) is produced in which individual points are connected by a line in the order of their appearance in the data file. Histograms representing the distribution of each variable are displayed along the diagonal of the matrix (in square matrices) or along the edges.

Matrix plots scatterplot In this type of Matrix Plot, 2D Scatterplots are arranged in a matrix format (values of the column variable are used as X coordinates, values of the row variable represent the Y coordinates). Histograms representing the distribution of each variable are displayed along the diagonal of the matrix or along the edges (in rectangular matrices).

Matrix rank The column (or row) rank of a rectangular matrix of values (e.g., a sums of squares and cross-products matrix) is equal to the number of linearly independent columns (or rows) of elements in the matrix. If there are no columns that are linearly dependent on other columns, then the rank of the matrix is equal to the number of its columns and the matrix is said to have full (column) rank. If the rank is less than the number of columns, the matrix is said to have reduced (column) rank and is singular.

Matrix singularity A rectangular matrix of values (e.g., a sums of squares and cross-products matrix) is singular if the elements in a column (or row) of the matrix are linearly dependent on the elements in one or more other columns (or rows) of the matrix. For example, if the elements in one column of a matrix are 1, -1, 0, and the elements in another column of the matrix are 2, -2, 0, then the matrix is singular because 2 times each of the elements in the first column is equal to each of the respective elements in the second column. Such matrices are also said to suffer from multicollinearity problems, since one or more columns are linearly related to each other.

A unique, regular matrix inverse cannot be computed for singular matrices, but generalized inverses (an infinite number of them) can be computed for any singular matrix.

See also matrix inverse.

Maximum likelihood estimate (MLE) The maximum likelihood estimate of a parameter from data is the possible value of the parameter for which the chance of observing the data largest. That is, suppose that the parameter is *p*, and that we observe data *x*. Then the maximum likelihood estimate of *p* is estimate *p* by the value *q* that makes P(observing *x* when the value of *p* is *q*) as large as possible.

For example, suppose we are trying to estimate the chance that a (possibly biased) coin lands heads when it is tossed. Our data will be the number of times *x* the coin lands heads in *n* independent tosses of the coin. The distribution of the number of times the coin lands heads is binomial with parameters *n* (known) and *p* (unknown). The chance of observing *x* heads in *n* trials if the chance of heads in a given trial is *q* is ${}_nC_x$ $q^x(1\text{-}q)^{n\text{-}x}$. The maximum likelihood estimate of *p* would be the value of *q* that makes that chance largest. We can find that value of *q* explicitly using calculus; it turns out to be $q = x/n$, the fraction of times the coin is observed to land heads in the *n* tosses. Thus the maximum likelihood estimate of the chance of heads from the number of heads in *n* independent tosses of the coin is the observed fraction of tosses in which the coin lands heads.

Maximum likelihood loss function An common alternative to the least squares loss function is to maximize the likelihood or log-likelihood function (or to minimize the negative log-likelihood function; the term maximum likelihood was first used by Fisher, 1922a). These functions are typically used when fitting non-linear models. In most general terms, the likelihood function is defined as:

$$L=F(Y,\text{Model})=\prod^{n}_{i=1} \{ p[y_i , \text{Model Parameters}(x_i)]\}$$

Maximum likelihood method The method of maximum likelihood (the term first used by Fisher, 1922a) is a general method of estimating parameters of a population by values that maximize the likelihood (L) of a sample. The likelihood L of a sample of n observations $x_1, x_2, ..., x_n$, is the joint probability function $p(x_1, x_2, ..., x_n)$ when x_1,

$x_2, ..., x_n$ are discrete random variables. If $x_1, x_2, ..., x_n$ are continuous random variables, then the likelihood L of a sample of n observations, $x_1, x_2, ..., x_n$, is the joint density function $f(x_1, x_2, ..., x_n)$.

Let L be the likelihood of a sample, where L is a function of the parameters $\theta_1, \theta_2, ... \theta_k$. Then the maximum likelihood estimators of $\theta_1, \theta_2, ... \theta_k$ are the values of $\theta_1, \theta_2, ... \theta_k$ that maximize L.

Let θbe an element of Ω. If Ω is an open interval, and if L(θ) is differentiable and assumes a maximum on W, then the MLE will be a solution of the following equation: $(dL(\theta))/d\theta = 0$.

Maximum unconfounding Maximum unconfounding is an experimental design criterion that is subsidiary to the criterion of design resolution. The maximum unconfounding criterion specifies that design generators should be chosen such that the maximum number of interactions of less than or equal to the crucial order, given the resolution, are unconfounded with all other interactions of the crucial order. It is an alternative to the minimum aberration criterion for finding the "best" design of maximum resolution.

Mcnemar's test A hypothesis test about the difference in success probabilities based on one i.i.d. sample of paired values of binary variables. Equivalent to a sign test on binary data.

MD (missing data) Same as Missing values.

Mean The mean is a particularly informative measure of the "central tendency" of the variable if it is reported along with its confidence intervals. Usually we are interested in statistics (such as the mean) from our sample only to the extent to which they are informative about the population. The larger the sample size, the more reliable its mean. The larger the variation of data values, the less reliable the mean.

$$\text{Mean} = (\Sigma x_i)/n$$

where n is the sample size.

Mean square In an analysis of variance table, the ratio of a sum-of-squares to its corresponding degrees of freedom.

Mean squared error (MSE) The mean squared error of an estimator of a parameter is the expected value of the square of the difference between the estimator and the parameter. In symbols, if X is an estimator of the parameter *t*, then

$$MSE(X) = E((X-t)^2).$$

The MSE measures how far the estimator is off from what it is trying to estimate, on the average in repeated experiments. It is a summary measure of the accuracy of the estimator. It combines any tendency of the estimator to overshoot or undershoot the truth (bias), and the variability of the estimator (SE). The MSE can be written in terms of the bias and SE of the estimator:

$$MSE(X) = (\text{bias}(X))^2 + (SE(X))^2.$$

Mean substitution of missing data When you select Mean Substitution, the missing data will be replaced by the means for the respective variables during an analysis.

Mean/S.D. An algorithm (used in neural networks) to assign linear scaling coefficients for a set of numbers. The mean and standard deviation of the set are found, and scaling factors selected so that these are mapped to desired mean and standard deviation values.

Measure The goal of the Measure phase is to gather information about the current situation, to obtain baseline data on current process perfòrmance, and to identify problem areas.

Measure of association A statistic which measures the degree of association or relationship between two variables. Examples include phi for nominal variables and the correlation coefficient for interval variables.

Measure of central tendency A single statistic that summarises the distribution of one variable. See mode, median and mean.

Measurement type This is a distinction regarding the relationship between a phenomenon being measured and the data as recorded. The main distinctions are concerned with the meaningfulness of numerical comparisons of data (nominal scale versus ordinal scale versus interval scale versus ratio scale : this is known as stevens' typology), whether the scale of the measurements (other than nominal scale measurements) should be regarded as essentially conituous or discrete, and whether the scale is bounded or unbounded.

Measures of central tendency Descriptive measures that indicate the center of a set of values, for example, mean, median, and mode.

Measures of dispersion Statistics that describes the spread of a distribution or how the values of a distribution are scattered around the mean. The simplest is the range, but see also variance and standard deviation.

Measures of variation Descriptive measures that indicate the dispersion of a set of values, for example, variance, standard deviation, and standard error of the mean.

Median The median is the value halfway through the ordered data set, below and above which there lies an equal number of data values.

Example

With an odd number of data values, for example 21, we have:

Data 96 48 27 72 39 70 7 68 99 36 95 4 6 13 34 74 65 42 28 54 69

Ordered Data4 6 7 13 27 28 34 36 39 42 48 54 65 68 69 70 72 74 95 96 99

Median 48, leaving 10 values below and 10 values above.

With an even number of data values, for example 20, we have:

Data 57 55 85 24 33 49 94 2 8 51 71 30 91 6 47 50 65 43 41 7.

Ordered Data2 6 7 8 24 30 33 41 43 47 49 50 51 55 57 65 71 85 91 94.

Median Halfway between the two 'middle' data points - in this case halfway between 47 and 49 = 48.

It is generally a good descriptive measure of the location which works well for skewed data, or data with outliers.

The median is the 0.5 quantile.

Member of a set Something is a member (or element) of a set if it is one of the things in the set.

Meta learning The concept of meta-learning applies to the area of predictive data mining, to combine the predictions from multiple models. It is particularly useful when the types of models included in the project are very different. In this context, this procedure is also referred to as Stacking (Stacked Generalization).

Suppose your data mining project includes tree classifiers, such as C&RT and chaid, linear discriminant analysis (e.g., GDA), and Neural Networks. Each computes predicted classifications for a crossvalidation sample, from which overall goodness-of-fit statistics (e.g., misclassification rates) can be computed. Experience has shown that combining the predictions from multiple methods often yields more accurate predictions than can be derived from any one method. The predictions from different classifiers can be used as input into a meta-learner, which will attempt to combine the predictions to create a final best predicted classification. So, for example, the predicted classifications from the tree classifiers, linear model, and the neural network classifier(s) can be used as input variables into a neural network meta-classifier, which will attempt to "learn" from the data how to combine the predictions from the different models to yield maximum classification accuracy.

One can apply meta-learners to the results from different meta-learners to create "meta-meta"-learners, and so on; however, in practice such exponential increase in the amount of data processing, in order to derive an accurate prediction, will yield less and less marginal utility.

Method of comparison The most basic and important method of determining whether a treatment has an effect: compare what happens to individuals who are treated (the treatment group) with what happens to individuals who are not treated (the control group).

MID P This is a tail definition policy that the alpha value should be calculated as the sum of the proportion of the tail for data strictly more extreme than the outcome, plus one half of the proportion of the distribution corresponding to the exact outcome value. This gives an unbiased estimate of alpha.

Middle quartile (Q2) The 50th percentile of a set of measures; the median.

Minimal change sequence Exploration of a randomisation distribution is such a sequence that the successive randomisations differ is a simple way. In the context of a ranodmisation test this can mean that the value of the test statistic for a particular randomisation may be calculated by a simple adjustment to the value for the preceding randomisation.

Minimax An algorithm to assign linear scaling coefficients for a set of numbers. The minimum and maximum of the set are found, and scaling factors selected so that these are mapped to desired minimum and maximum values.

Minimax strategy In game theory, a minimax strategy is one that minimizes one's maximum loss, whatever the opponent might do (whatever strategy the opponent might choose).

Minimum aberration Minimum aberration is an experimental design criterion that is subsidiary to the criterion of design resolution. The minimum aberration design is defined as the design of maximum resolution "which minimizes the number of words in the defining relation that are of minimum length" (Fries & Hunter, 1980). Less technically, the criterion apparently operates by choosing design generators that produce the smallest number of pairs of confounded interactions of the crucial order. For example, the minimum aberration resolution IV design would have the minimum number of pairs of confounded 2-factor interactions.

Minimum improvement Specifies that the RMS error on the training subset, the selection subset, or both must improve by at least this amount, or training will cease (if the Window parameter is non-zero).

Sometimes error improvement may slow down for a while or even rise temporarily (particularly if the shuffle option is used with back propagation, or non-zero noise is specified, as these both introduce an element of noise into the training process).

To prevent this option from aborting the run prematurely, specify a longer Window.

It is particularly recommended to monitor the selection error for minimum improvement, as this helps to prevent over-learning.

Specify a negative improvement threshold if you want to stop training only when a significant deterioration in the error is detected. The algorithm will stop when a number of generations pass during which the error is always the given amount worse than the best it ever achieved.

Missing values It is not uncommon to come across situations where the data for some cases has some values missing; perhaps because data was unavailable, or corrupted, when gathered. In such cases, you may still need to execute a network (to get the best estimate possible given the information available) or (and this is more suspect) use the partially complete data in training because of an acute shortage of training data.

Where possible, it is usually good practice not to use variables containing a great many missing values. Cases with missing values can be excluded.

Mode Most frequent measure in a distribution; the high point on a frequency distribution.

Mode The mode is the most frequently occurring value in a set of discrete data. There can be more than one mode if two or more values are equally common.

Example

Suppose the results of an end of term Statistics exam were distributed as follows:

Student:.....Score:

1.............94

2.............81

3.............56

4.............90

5.............70

6.............65

7.............90

8.............90

9.............30

Then the mode (most common score) is 90, and the median (middle score) is 81.

Model A theory that proposes relationships between two or more variables.

Model profiles Model profiles are concise text strings indicating the architecture of networks and ensembles. A profile consists of a type code followed by a code giving the number of input and output variables and number of layers and units (networks) or members (ensembles). For time series networks, the number of steps and the lookahead factor are also given.

Models for data mining In the business environment, complex data mining projects may require the coordinate efforts of various experts, stakeholders, or departments throughout an entire organization. In the data mining literature, various "general frameworks" have been proposed to serve as blueprints for how to organize the process of gathering data, analyzing data, disseminating results, implementing results, and monitoring improvements.

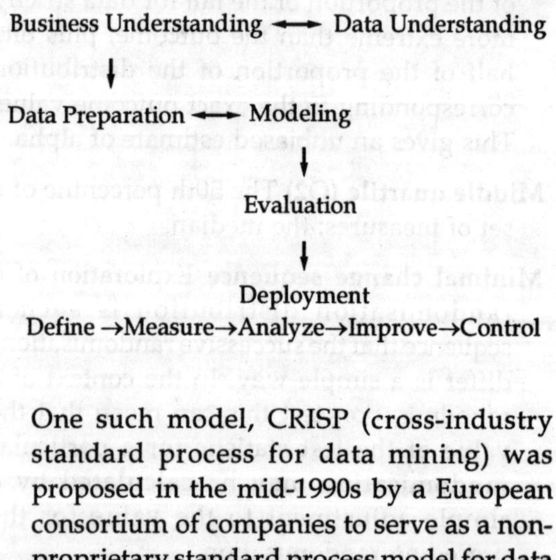

Define →Measure→Analyze→Improve→Control

One such model, CRISP (cross-industry standard process for data mining) was proposed in the mid-1990s by a European consortium of companies to serve as a non-proprietary standard process model for data mining. This general approach postulates the following (perhaps not particularly controversial) general sequence of steps for data mining projects:

Another approach - the Six Sigma methodology - is a well-structured, data-driven methodology for eliminating defects, waste, or quality control problems of all kinds in manufacturing, service delivery, management, and other business activities. This model has recently become very popular (due to its successful implementations) in various American industries, and it appears to gain favor worldwide. It postulated a sequence of, so-called, DMAIC steps -

- that grew up from the manufacturing, quality improvement, and process control traditions and is particularly well suited to production environments (including "production of services," i.e., service industries).

Another framework of this kind (actually somewhat similar to Six Sigma) is the approach proposed by SAS Institute called SEMMA -

which is focusing more on the technical activities typically involved in a data mining project.

All of these models are concerned with the process of how to integrate data mining methodology into an organization, how to "convert data into information," how to involve important stake-holders, and how to disseminate the information in a form that can easily be converted by stake-holders into resources for strategic decision making.

Some software tools for data mining are specifically designed and documented to fit into one of these specific frameworks.

The general underlying philosophy of StatSoft's Statistica Data Miner is to provide a flexible data mining workbench that can be integrated into any organization, industry, or organizational culture, regardless of the general data mining process-model that the organization chooses to adopt. For example, Statistica Data Miner can include the complete set of (specific) necessary tools for ongoing company wide Six Sigma quality control efforts, and users can take advantage of its (still optional) DMAIC-centric user interface for industrial data mining tools. It can equally well be integrated into ongoing marketing research, CRM (Customer Relationship Management) projects, etc. that follow either the CRISP or SEMMA approach - it fits both of them perfectly well without favoring either one. Also, Statistica Data Miner offers all the advantages of a general data mining oriented "development kit" that includes easy to use tools for incorporating into your projects not only such components as custom database gateway solutions, prompted interactive queries, or proprietary algorithms, but also systems of access privileges, workgroup management, and other collaborative work tools that allow you to design large scale, enterprise-wide systems (e.g., following the CRIPS, SEMMA, or a combination of both models) that involve your entire organization.

Moment The *k*th moment of a list is the average value of the elements raised to the *k*th power; that is, if the list consists of the *N* elements $x_1, x_2, \ldots, x_N$, the *k*th moment of the list is

$(x_1^k + x_2^k + x_N^k)/N$.

The *k*th moment of a random variable X is the expected value of X^k, $E(X^k)$.

Monotone, monotonic function A function is monotone if it only increases or only decreases: *f* increases monotonically (is monotonic increasing) if $x > y$, implies that$f(x) >= f(y)$. A function *f* decreases monotonically (is monotonic decreasing) if $x > y$, implies that$f(x) <= f(y)$. A function *f* is strictly monotonically increasing if $x > y$, implies that$f(x) > f(y)$, and strictly monotonically decreasing if if $x > y$, implies that$f(x) < f(y)$.

Monte Carlo A computer-intensive technique for assessing how a statistic will perform under repeated sampling. In Monte Carlo methods, the computer uses random number simulation techniques to mimic a statistical population. In the Statistica Monte Carlo procedure, the computer constructs the population according to the user's prescription, then does the following:

For each Monte Carlo replication, the computer:

Simulates a random sample from the population,

Analyzes the sample,

Stores the results.

After many replications, the stored results will mimic the sampling distribution of the statistic. Monte Carlo techniques can provide information about sampling distributions when exact theory for the sampling distribution is not available.

Monte Carlo test [Named after the famous site of gambling casinos] A monte-carlo test involves generating a random subset of the randomisation set, sampled without replacement, and using the values of the test statistic to generate an estimate of the form of the full randomisation distribution. This procedure is in contrast to the bootstrap procedure in that the sampling of the randomisation set is without replacement. An advantage of the monte-carlo test over the bootstrap is that with successive resamplings it converges to the gold standard(1) form of the exact test(1). An effective necessity for the monte-carlo procedure is a source of random codes or an effective pseudo-random generator.

Mound shaped curve Symmetrical, single-peaked frequency distribution. Also called the normal curve or gaussian curve. Also called a bell-shaped curve.

Moving average smoothing A moving average is a form of average which has been adjusted to allow for seasonal or cyclical components of a time series. Moving average smoothing is a smoothing technique used to make the long term trends of a time series clearer.

When a variable, like the number of unemployed, or the cost of strawberries, is graphed against time, there are likely to be considerable seasonal or cyclical components in the variation. These may make it difficult to see the underlying trend. These components can be eliminated by taking a suitable moving average.

By reducing random fluctuations, moving average smoothing makes long term trends clearer.

Mpatt bar Multi-pattern bar plots may be used to represent individual data values of the X variable (the same type of data as in pie charts), however, consecutive data values of the X variable are represented by the heights of sequential vertical bars, each of a different color and pattern (rather than as pie wedges of different widths).

Multi dimensional scaling Multidimensional scaling (MDS) can be considered to be an alternative to factor analysis (see Factor Analysis), and it is typically used as an exploratory method. In general, the goal of the analysis is to detect meaningful underlying dimensions that allow the researcher to explain observed similarities or dissimilarities (distances) between the investigated objects. In factor analysis, the similarities between objects (e.g., variables) are expressed in the correlation matrix. With MDS one may analyze not only correlation matrices but also any kind of similarity or dissimilarity matrix (including sets of measures that are not internally consistent, e.g., do not follow the rule of transitivity).

Multimodal distribution A distribution that has multiple modes (thus two or more "peaks").

Multimodality of the distribution in a sample is often a strong indication that the distribution of the variable in population is not normal. Multimodality of the distribution may provide important information about the nature of the investigated variable (i.e., the measured quality). For example, if the variable represents a reported preference or attitude, then multimodality may indicate that there are several pronounced views or patterns of response in the questionnaire. Often however, the multimodality may indicate that the sample is not homogenous and the observations come in fact from two or more "overlapping" distributions. Sometimes, multimodality of the distribution may indicate problems with the measurement instrument (e.g, "gage calibration problems" in natural sciences, or "response biases" in social sciences).

See also unimodal distribution, bimodal distribution.

Multilayer perceptrons Feedforward neural networks having linear PSP functions and (usually) non-linear activation functions.

Multi variate analysis Statistical analysis involving three or more variables.

Multinomial distribution The multinomial distribution arises when a response variable is categorical in nature, i.e., consists of data describing the membership of the respective cases to a particular category. For example, if a researcher recorded the outcome for the driver in accidents as "uninjured, "injury not requiring hospitalization", "injury requiring hospitalization", or "fatality", then the distribution of the counts in these categories would be multinomial. The multinomial distribution is a generalization of the binomial distribution to more than two categories.

If the categories for the response variable can be ordered, then the distribution of that variable is referred to as ordinal multinomial. For example, if in a survey the responses to a question are recorded such that respondents have to choose from the pre-arranged categories "Strongly agree", "Agree", "Neither agree nor disagree", "Disagree", and "Strongly disagree", then the counts (number of respondents) that endorsed the different categories would follow an ordinal multinomial distribution (since the response categories are ordered with respect to increasing degrees of disagreement.

Multinomial logit and probit regression The multinomial logit and probit regression models are extensions of the standard logit and probit regression models to the case where the dependent variable has more than two categories (e.g., not just Pass - Fail, but Pass, Fail, Withdrawn), i.e., when the dependent or response variable of interest follows a multinomial distribution rather than binomial distribution. When multinomial responses contain rank-order information, they are also called ordinal multinomial responses.

Multiple axes in graphs An arrangement of axes (coordinate scales) in graphs, where two or more axes are placed parallel to each other, in order to either:

represent different units in which the variable(s) depicted in the graph can be measured (e.g., a Celsius and Fahrenheit scales of temperature), or

allow for a comparison of trends or shapes between several plots placed in one graph (e.g., one axis for each plot) which otherwise would be obscured by incompatible measurement units or ranges of values for each variable (that is an extension of the common "double-Y" type of graph).

Multiple dichotomies One possible coding scheme that can be used when more than one response is possible from a given question is to code responses using Multiple dichotomies . For example, as part of a larger market survey, suppose you asked a sample of consumers to name their three favorite soft drinks. The specific item on the questionnaire may look like this:

Write down your three favorite soft drinks 1:_________ 2:_________ 3:_________

Suppose in the above example we were only interested in Coke, Pepsi, and Sprite. One way to code the data in that case would be as follows:

coke pepsi sprite . . . .

case 1case 2 case 3 . . . 1 . . 1 1 .

. . 1 . . .

In other words, one variable was created for each soft drink, then a value of 1 was entered into the respective variable whenever the respective drink was mentioned by the respective respondent. Note that each variable represents a dichotomy; that is, only "1"s and "not 1"s are allowed (we could have entered 1's and 0's, but to save typing we can also simply leave the 0's as blanks or as missing values). When tabulating these variables, we would like to compute the number and percent of respondents (and responses) for each soft drink. In a sense, we "compact" the three variables Coke, Pepsi, and Sprite into a single variable (Soft Drink) consisting of multiple dichotomies.

Multiple histogram Multiple histograms present frequency distributions of more than one variable in one 2D graph. Unlike the Double-Y Histograms, the frequencies for all variables are plotted against the same left-Y axis.

Also, the values of all examined variables are plotted against a single X-axis, which facilitates comparisons between analyzed variables.

Multiple output neurons In many-class problems, the target output is 1.0 in the correct class output and 0.0 in the others.

The output neuron activation levels provide confidence estimates for the output classes. It is desirable to be able to interpret these confidence levels as probabilities. If the correct choice of network error function is used during optimization, combined with the correct activation function, such as interpretation can be made. Specifically, a cross entropy error function is combined with the logistic activation function for a two-class problem encoded by a single output neuron or with softmax for a three or more class problem.

The entropic approach corresponds to maximum likelihood optimization, assuming that the data is drawn from the exponential family of distributions. An important feature is that the outputs may be interpreted as posterior estimates of class membership probability.

The alternative approach is to use a sum-squared error function with the logistic output activation function. This has less statistical justification - the network learns a discriminant function, and although the outputs can be treated as confidence measures, they are not probability estimates (and indeed may not even sum to 1.0). On the other hand, such networks sometimes train more quickly, the training process is more stable, and they may achieve higher classification rates. If no thresholds are used, the network uses a "winner takes all" algorithm; the highest activation neuron gives the class. There is a "no doubt" option.

If you specify thresholds, the class is still assigned to the highest neuron, but there is a doubt option. The highest neuron's activation must be above the accept threshold and all other neuron's below the reject threshold in order for the class to be assigned; if this condition is not fulfilled, the class is "unknown."

If your multiple output neuron classification network is using the softmax activation function, the output neuron activations are guaranteed to sum to 1.0 and can be interpreted as probabilities of class membership. However, in other cases, although the activations can be interpreted as confidence levels in some sense (i.e. a higher number indicates greater confidence), they are not probabilities, and should be interpreted with caution.

Multiple R The coefficient of multiple correlation (Multiple R) is the positive square root of R-square (the coefficient of multiple determination). This statistic is useful in multivariate regression (i.e., multiple independent variables) when you want to describe the relationship between the variables.

Multiple regression The general purpose of multiple regression (the term was first used by Pearson, 1908) is to analyze the relationship between several independent or predictor variables and a dependent or criterion variable.

The computational problem that needs to be solved in multiple regression analysis is to fit a straight line (or plane in an n-dimensional space, where n is the number of independent variables) to a number of points. In the simplest case – one dependent and one independent variable – one can visualize this in a scatterplot (scatterplots are two-dimensional plots of the scores on a pair of variables). It is used as either a hypothesis testing or exploratory method.

Multiple regression analysis Statistical methods concerned with explaining or predicting the variability of a continuous dependent variable using information from two or more continuous independent variables.

Multiple regression correlation coefficient The multiple regression correlation coefficient, R^2, is a measure of the proportion of variability explained by, or due to the regression (linear relationship) in a sample of paired data. It is a number between zero and one and a value close to zero suggests a poor model.

A very high value of R^2 can arise even though the relationship between the two variables is non-linear. The fit of a model should never simply be judged from the R^2 value.

Multiple-response tables Multiple-response tables are Crosstabulation tables used when the categories of interest are not mutually exclusive. Such tables can accommodate Multiple response variables as well as Multiple dichotomies.

Multiple response variables Coding the responses to Multiple response variables is necessary when more than one response is possible from a given question. For example, as part of a larger market survey, suppose you asked a sample of consumers to name their three favorite soft drinks. The specific item on the questionnaire may look like this:

Write down your three favorite soft drinks
1:__________ 2:__________ 3:__________

Thus, the questionnaires returned to you will contain somewhere between 0 and 3 answers to this item. Also, a wide variety of soft drinks will most likely be named. One way to record the various responses would be to use three multiple response variables and a coding scheme for the many soft drinks. Then we could enter the respective codes (or alphanumeric labels) into the three variables, in the same way that respondents wrote them down in the questionnaire.

Multiple stream group charts Variable and attribute control charts (see also Quality Control) can be computed for multiple-stream processes (e.g., operators, machines, assembly lines); the resulting multiple stream group chart summarizes the measurements for all streams simultaneously. These charts can also be produced for short production runs, and the measurements summarized in short run group charts. In addition to the standard parameters for determining the control limits and other characteristics of the control charts, the number of consecutive points r from the same process stream (i.e., "runs" of length r) to be highlighted in the chart can be specified.

Multiplication rule The multiplication rule is a result used to determine the probability that two events, A and B, both occur.

The multiplication rule follows from the definition of conditional probability.

The result is often written as follows, using set notation:

$P(A \cap B) = P(A|B).P(B)$ or $P(A \cap B) = P(B|A).P(A)$

where:

$P(A)$ = probability that event A occurs.

$P(B)$ = probability that event B occurs.

$P(A \cap B)$ = probability that event A and event B occur.

$P(A | B)$ = the conditional probability that event A occurs given that event B has occurred already.

$P(B | A)$ = the conditional probability that event B occurs given that event A has occurred already.

For independent events, that is events which have no influence on one another, the rule simplifies to:

$P(A \cap B) = P(A).P(B)$

That is, the probability of the joint events A and B is equal to the product of the individual probabilities for the two events.

Multiplicative season, damped trend In this Time Series model, the simple exponential smoothing forecasts are "enhanced" both by a damped trend component (independently smoothed with the single parameter ϕ; this model is an extension of Brown's one-parameter linear model) and a multiplicative seasonal component (smoothed with parameter δ). For example, suppose we wanted to forecast from month to month the number of households that purchase a particular consumer electronics device (e.g., VCR). Every year, the number of households that purchase a VCR will increase, however, this trend will be damped (i.e., the upward trend will slowly disappear) over time as the market becomes saturated. In addition, there will be a seasonal component, reflecting the seasonal changes in consumer demand for VCR's from month to month (demand will likely be smaller in the summer and greater during the December holidays). This seasonal component may be multiplicative, for example, sales during the December holidays may increase by factor of 1.4 (or 40%) over the average annual sales. To compute the smoothed values for the first season, initial values for the seasonal components are necessary. Also, to compute the smoothed value (forecast) for the first observation in the series, both estimates of S_0 and T_0 (initial trend) are necessary. These values are computed as:

$T_0 = (1/\phi)*M_k-M_1)/[(k-1)*p]$

where ϕ is the smoothing parameter

k is the number of complete seasonal cycles

M_k is the mean for the last seasonal cycle

M_1 is the mean for the first seasonal cycle

p is the length of the seasonal cycle

and $S_0 = M_1-p*T_0/2$.

Multiplicative season, exponential trend In this Time Series model, the simple exponential smoothing forecasts are "enhanced" both by an exponential trend component (independently smoothed with parameter γ) and a multiplicative seasonal component (smoothed with parameter δ). For example, suppose we wanted to forecast the monthly revenue for a resort area. Every year, revenue may increase by a certain percentage or factor, resulting in an exponential trend in overall revenue. In addition, there could be an multiplicative seasonal component, that is, given the respective annual revenue, each year 20% of the revenue is produced during the month of December, that is, during Decembers the revenue grows by a particular (multiplicative) factor.

To compute the smoothed values for the first season, initial values for the seasonal components are necessary. Also, to compute the smoothed value (forecast) for the first observation in the series, both estimates of S_0 and T_0 (initial trend) are necessary. By default, these values are computed as:

$T_0 = exp\{[log(M_2)-log(M_1)]/p\}$

where M_2 is the mean for the second seasonal cycle

M_1 is the mean for the first seasonal cycle

p is the length of the seasonal cycle

and $S_0 = exp\{log(M_1)-p^*log(T_0)/2\}$

Multiplicative season, linear trend In this Time Series model, the simple exponential smoothing forecasts are "enhanced" both by a linear trend component (independently smoothed with parameter γ) and a multiplicative seasonal component (smoothed with parameter δ). For example, suppose we were to predict the monthly budget for snow-removal in a community. There may be a trend component (as the community grows, there is an upward trend for the cost of snow removal from year to year). At the same time, there is obviously a seasonal component, reflecting the differential likelihood of snow during different months of the year. This seasonal component could be multiplicative, meaning that given a respective budget figure, it may increase by a factor of, for example, 1.4 during particular winter months; or it may be additive, that is, a particular fixed additional amount of money is necessary during the winter months. To compute the smoothed values for the first season, initial values for the seasonal components are necessary. Also, to compute the smoothed value (forecast) for the first observation in the series, both estimates of S_0 and T_0 (initial trend) are necessary. By default, these values are computed as:

$T_0 = (M_k-M_1)/((k-1)^*p)$

where k is the number of complete seasonal cycles

M_k is the mean for the last seasonal cycle

M_1 is the mean for the first seasonal cycle

p is the length of the seasonal cycle

and $S_0 = M_1 - T_0/2$.

Multiplicative season, no trend This Time Series model is partially equivalent to the simple exponential smoothing model; however, in addition, each forecast is "enhanced" by a multiplicative component that is smoothed independently. This model would, for example, be adequate when computing forecasts for monthly expected sales for a particular toy. The level of sales may be stable from year to year, or change only slowly; at the same time, there will be seasonal changes (e.g., greater sales during the December holidays), which again may change slowly from year to year. The seasonal changes may affect the sales in a multiplicative fashion, for example, depending on the respective overall level of sales, December sales may always be greater by a factor of 1.4.

Multiplicity in hypothesis tests In hypothesis testing, if more than one hypothesis is tested, the actual significance level of the combined tests is not equal to the nominal significance level of the individual tests. See also false discovery rate.

Multivariate adaptive regression splines (MARsplines) Multivariate adaptive regression splines (or MARSplines for short) is a nonparametric regression procedure which makes no assumption about the underlying functional relationship between the dependent and independent variables. Instead MARSplines constructs this relation from a set of coefficients and basis functions that are entirely "driven" from the regression data. The MARSplines technique has become particularly popular in the area of data mining, because it does not assume or impose any particular type or class of relationship (e.g., linear, logistic, and so on) between the predictor variables and the dependent (outcome) variable of interest.

The general MARSplines model equation is given as:

$$y = f(X) = \beta_0 + \sum_{m-1}^{M} \beta_m h_m(X)$$

where the summation is over the M predictors in the model. To summarize, y is predicted as a function of the predictor variables X (and their interactions); this function consists of an intercept parameter (β_0 and the weighted (by β_m sum of one or more basis functions $h_m(X)$. You may also think of this model as "selecting" a weighted sum of basis functions from the set of (a large number of) basis functions that span all values of each predictor (i.e., that set would consist of one basis function, and "knot" parameter t, for each distinct value for each predictor variable): The MARSplines algorithm then searches over the space of all inputs and predictor values (knot locations t) as well as interactions between variables. During this search an increasingly larger number of basis functions are added to the model (selected from the set of possible basis functions), to maximize an overall least squares goodness-of-fit criterion.

Multivariate data A set of measurements of two or more variables per individual. See bivariate.

Mutually exclusive (also called (pairwise) disjoint) Of two events, the property that they cannot both happen; of more than two events, the property that the intersection of any pair is impossible.

Mutually exclusive Events such that the occurrence of one precludes the occurrence of the other.

Mutually exclusive events Two events are mutually exclusive (or disjoint) if it is impossible for them to occur together.

Formally, two events A and B are mutually exclusive if and only if :

$A \cap B = \phi$

If two events are mutually exclusive, they cannot be independent and vice versa.

Examples

1. Experiment Rolling a die once.

Sample space S = {1,2,3,4,5,6}.

Events A = 'observe an odd number' = {1,3,5}.

B = 'observe an even number' = {2,4,6}.

$A \cap B = \phi$ = the empty set, so A and B are mutually exclusive.

2. A subject in a study cannot be both male and female, nor can they be aged 20 and 30. A subject could however be both male and 20, both female and 30.

N

N in one encoding For nominal variables with more than two states, the practice of representing the variable using a single unit with a range of possible values (actually implemented using minimax, explicit or none).

N point moving average line Each point on this moving average line represents the average of the respective sample and the n-1 number of preceding samples. Thus, this line will smooth the pattern of means across samples, allowing the quality control engineer to detect trends. You can specify the number of samples (n) that are to be averaged for each point in the plot.

Naive bayes A statistical method based on Bayesian theorem that is primarily used for classification tasks.

Nearly normal distribution A population of numbers (a list of numbers) is said to have a *nearly normal distribution* if the histogram of its values in standard units nearly follows a normal curve. More precisely, suppose that the mean of the list is μ and the standard deviation of the list is SD. Then the list is nearly normally distributed if, for every two numbers $a < b$, the fraction of numbers in the list that are between a and b is approximately equal to the area under the normal curve between $(a - \mu)/SD$ and $(a - \mu)/SD$.

Neat scaling of intervals The term neat scaling is used to refer to the manner in which ranges of values are divided into intervals, so that the resulting interval boundaries and steps between those boundaries are intuitive and readily interpretable (or "understood").

For example, suppose you want to create a histogram for data values in the range from 1 to 10. It would be inefficient to use interval boundaries for the histogram at values such as 1.3, 3.9, 6.5, etc., i.e., to use as a minimum boundary value 1.3, and then a step size of 2.6. A much more intuitive way to divide the range of data values would be to use boundaries like 1, 2, 3, 4, and so on, i.e., a minimum boundary at 1, with step size of 1; or one could use 2, 4, 6, etc, i.e., a minimum boundary of 2 and step size 2.

In general, neat in this context means that category boundaries will be round values ending either in 0, 2, or 5 (e.g., boundaries may be 0.1, 0.2, 0.3, etc.; or 50, 100, 150, etc.). To achieve this, any user-requested lower limit, upper limit, and number of categories will only be approximated.

Negative binomial distribution Consider a sequence of independent trials with the same probability p of success in each trial. The number of trials up to and including the rth success has the negative Binomial distribution with parameters n and r. If the random variable N has the negative binomial distribution with parameters n and r, then

$$P(N=k) = {}_{k-1}C_{r-1} \times p^r \times (1-p)^{k-r},$$

for $k = r, r+1, r+2, \ldots,$ and zero for $k < r$, because there must be at least r trials to have r successes. The negative binomial distribution is derived as follows: for the rth success to occur on the kth trial, there must have been r-1 successes in the first k-1 trials, and the kth trial must result in success. The chance of the former is the chance of r-1 successes in k-1 independent trials with the same probability of success in each trial, which, according to the Binomial distribution with parameters $n=k-1$ and p, has probability

$${}_{k-1}C_{r-1} \times p^{r-1} \times (1-p)^{k-r}.$$

The chance of the latter event is p, by assumption. Because the trials are independent, we can find the chance that both events occur by multiplying their chances together, which gives the expression for $P(N=k)$ above.

Negative correlation The relationship between two variables is such that as one variable's values tend to increase, the other variable's values tend to decrease. This is represented by a negative correlation coefficient.

Negative exponential (2D graphs) A curve is fitted to the XY coordinate data according to the negative exponentially-weighted smoothing procedure (the influence of individual points decreases exponentially with the horizontal distance from the respective points on the curve).

Negative exponential (3D graphs) A surface is fitted to the XYZ coordinate data according to the negative exponentially- weighted smoothing procedure (the influence of individual points decreases exponentially with the horizontal distance from the respective points on the surface).

Negative predictive value For a diagnostic procedure, the conditional probability of absence of disease given a negative test result.

Negative relationship A relationship between two variables such that when one increases, the other decreases.

Negative skew A graph that has scores bunching up toward the positive end of the abscissa.

Negatively skewed curve A probability or frequency distribution that is not normal, but rather is shifted such that the mean is less than the mode.

Negatively skewed distribution A distribution in which a small proportion of individuals have excessively small values. Negatively skewed distributions are characterized by having a "long left tail". For such distributions, the mean tends to be smaller than the median.

Neighborhood In Kohonen training, a square set of units focused around the "winning" unit and simultaneously updated by the training algorithm.

Nested collection An ordered collection of events such that each event is implied by any of those following it in the collection.

Nested factors In nested designs the levels of a factor are nested (the term was first used by Ganguli, 1941) within the levels of another factor. For example, if one were to administer four different tests to four high school classes (i.e., a between-groups factor with 4 levels), and two of those four classes are in high school A, whereas the other two classes are in high school B, then the levels of the first factor (4 different tests) would be nested in the second factor (2 different high schools).

Nested sequence of models In Structural Equation Modeling, a set of models M(1), M(2), ... M(k) form a nested sequence if model M(i) is a special case of M(i+1) for i=1 to k-1. Thus, each model in the sequence becomes increasingly more general, but includes all previous models as special cases. As an example, consider one factor, two factor, and three factor models for 10 variables. The two factor model includes the one factor model as a special case (simply let all the loadings on the second factor be 0). Similarly, the three factor model contains the two and one factor models as special cases.

Network architecture This is of the form I:N-N-N:O, where I is the number of input variable, O the number of output variables, and N the number of units in each layer.

Example 2:4-6-3:1 indicates a network with 2 input variables, 1 output variable, 4 input neurons, 6 hidden neurons, and 3 output neurons.

For a time series network, the steps factor is prepended to the profile, and signified by an "s."

Example s10 1:10-2-1:1 indicates a time series network with steps factor (lagged input) 10.

Neural networks Neural Networks are analytic techniques modeled after the (hypothesized) processes of learning in the cognitive system and the neurological functions of the brain and capable of predicting new observations (on specific variables) from other observations (on the same or other variables) after executing a process of so-called learning from existing data.

Neuron A unit in a neural network.

Newman Keuls test This post hoc test can be used to determine the significant differences between group means in an analysis of variance setting. The Newman-Keuls test, like Duncan's test, is based on the range statistic.

No causation without manipulation A slogan attributed to Paul Holland. If the conditions were not deliberately manipulated (for example, if the situation is an observational study rather than an experiment), it is unwise to conclude that there is any causal relationship between the outcome and the conditions.

Noise addition A practice (used in neural networks) designed to prevent overlearning during back propagation training, by adding random noise to input patterns during training (and so "blurring" the position of the training data).

Nominal A scale using numbers, symbols, or names to designate different subclasses.

Nominal alpha criterion level A publicly agreed value for type-1 error, such that the outcome of a statistical test is classified in terms of whether the obtained value of alpha is extreme as compared with this criterion level. The fine detail of the comparison involves the tail definition policy. The outcome is classified as showing statistical significance ('significant') if the outcome has low alpha as compared with the nominal alpha criterion level, otherwise not ('non-significant'). The commonest conventional values for the nominal alpha criterion level are 0.05 and 0.01 .

Nominal data A set of data is said to be nominal if the values / observations belonging to it can be assigned a code in the form of a number where the numbers are simply labels. You can count but not order or measure nominal data. For example, in a data set males could be coded as 0, females as 1; marital status of an individual could be coded as Y if married, N if single.

Nominal scale This is a categorical (i.e., quantative and not qualitative) scale of measurement where each value represents a specific category that the variable's values fall into (each category is "different" than the others but cannot be quantitatively compared to the others).

Nominal variable The categories of a nominal variable bear no relationship to one another, other than that they are different. For example, marital status is a nominal variable with the categories, married, single, divorced etc. Nominal variables are measured at the lowest level of measurement.

Non centering correction (K) We can correct C_p for the effects of non-centering. Specifically, we can compute:

K = abs(Target Specification - Mean)/(1/2(USL-LSL))

This correction factor expresses the non-centering (target specification minus mean) relative to the specification range. Demonstrated excellence (C_{pk}). Finally, we can adjust C_p for the effect of non-centering by computing:

$C_{pk} = (1\text{-}k)*C_p$

If the process is perfectly centered, then k is equal to zero, and C_{pk} is equal to C_p. However, as the process drifts from the target specification, k increases and C_{pk} becomes smaller than C_p.

Non directional test A test of the prediction that two values are equal or a test that they are not equal; a two-tailed test.

Non outlier range The non-outlier range is the range of values in the 2D Box Plots, 3D Sequential Graphs - Box Plots, or Categorized Box Plots, which fall below the upper outlier limit (for example, +1.5 * the height of the box) and above the lower outlier limit (for example, -1.5 * the height of the box).

Non parametric tests Non-Parametric tests are often used in place of their parametric counterparts when certain assumptions about the underlying population are questionable. For example, when comparing two independent samples, the Wilcoxon Mann-Whitney test does not assume that the difference between the samples is normally distributed whereas its parametric counterpart, the two sample t-test does. Non-Parametric tests may be, and often are, more powerful in detecting population differences when certain assumptions are not satisfied.

All tests involving ranked data, i.e. data that can be put in order, are non-parametric. A number of statistical tests were devised, mostly over the period 1930-1960, with the specific objective of by-passing assumptions about sampling from populations with data supposedly conforming to theoretically modelled statistical distributions wuch as the normal distribution. Several of these tests were explictly concerned with ordinal-scale data for which modelling based upon continuous functions is clearly inappropriate. These tests are implicitly re-randomisation tests.

Nonlinear association The relationship between two variables is nonlinear if a change in one is associated with a change in the other that is depends on the value of the first; that is, if the change in the second is not simply proportional to the change in the first, independent of the value of the first variable.

Nonlinear estimation In the most general terms, Nonlinear estimation involves finding the best fitting relationship between the values of a dependent variable and the values of a set of one or more independent variables (it is used as either a hypothesis testing or exploratory method). For example, we may want to compute the relationship between the dose of a drug and its effectiveness, the relationship between training and subsequent performance on a task, the relationship between the price of a house and the time it takes to sell it, etc. Research issues in these examples are commonly addressed by such techniques as multiple regression or analysis of variance. In fact, one may think of Nonlinear estimation as a generalization of those methods. Specifically, multiple regression (and ANOVA) assumes that the relationship between the independent variable(s) and the dependent variable is linear in nature. Nonlinear Estimation leaves it up to you to specify the nature of the relationship; for example, you may specify the dependent variable to be a logarithmic function of the independent variable(s), an exponential function, a function of some complex ratio of independent measures, etc. (However, if all variables of interest are categorical in nature, or can be converted into categorical variables, you may also consider Correspondence Analysis as an alternative analysis technique.).

Nonlinear regression Nonlinear regression aims to describe the relationship between a response variable and one or more explanatory variables in a non-linear fashion.

Nonparametrics Nonparametric methods were developed to be used in cases when the researcher does not know the parameters of the distribution of the variable of interest in the population (hence the name nonparametric). In more technical terms, nonparametric methods do not rely on the estimation of parameters (such as the mean or the standard deviation) describing the distribution of the variable of interest in the population. Therefore, these methods are also sometimes (and more appropriately) called parameter-free methods or distribution-free methods.

Nonresponse In surveys, it is rare that everyone who is "invited" to participate (everyone whose phone number is called, everyone who is mailed a questionnaire, everyone an interviewer tries to stop on the street . . .) in fact responds. The difference between the "invited" sample sought, and that obtained, is the nonresponse.

Nonresponse bias In a survey, those who respond may differ from those who do not, in ways that are related to the effect one is trying to measure. For example, a telephone survey of how many hours people work is likely to miss people who are working late, and are therefore not at home to answer the phone. When that happens, the survey may suffer from nonresponse bias. Nonresponse bias makes the result of a survey differ systematically from the truth.

Nonresponse rate The fraction of nonresponders in a survey: the number of nonresponders divided by the number of people invited to participate (the number sent questionnaires, the number of interview attempts, etc.) If the nonresponse rate is appreciable, the survey suffer from large nonresponse bias.

Nonseasonal, damped trend In this Time Series model, the simple exponential smoothing forecasts are "enhanced" by a damped trend component (independently smoothed with parameters γ for the trend, and ϕ for the damping effect). For example, suppose we wanted to forecast from month to month the percentage of households that own a particular consumer electronics device (e.g., a VCR). Every year, the proportion of households owning a VCR will increase, however, this trend will be damped (i.e., the upward trend will slowly disappear) over time as the market becomes saturated.

To compute the smoothed value (forecast) for the first observation in the series, both estimates of S_0 and T_0 (initial trend) are necessary. By default, these values are computed as:

$T_0 = (1/\phi)*(X_n-X_1)/(N-1)$

where

N is the number of cases in the series,

ϕ is the smoothing parameter for the damped trend

and $S_0 = X_1-T_0/2$.

Nonseasonal, exponential trend In this Time Series model, the simple exponential smoothing forecasts are "enhanced" by an exponential trend component (smoothed with parameter γ). For example, suppose we wanted to predict the overall monthly costs of repairs to a production facility. There could be an exponential trend in the cost, that is, from year to year the costs of repairs may increase by a certain percentage or factor, resulting in a gradual exponential increase in the absolute dollar costs of repairs.

To compute the smoothed value (forecast) for the first observation in the series, both estimates of S_0 and T_0 (initial trend) are necessary. By default, these values are computed as:

$T_0 = (X_2/X_1)$

and

$S_0 = X_1/T_0^{1/2}$.

Nonseasonal, linear trend In this Time Series model, the simple exponential smoothing forecasts are "enhanced" by a linear trend component that is smoothed independently via the γ (gamma) parameter. This model is also referred to as Holt's two parameter method. This model would, for example, be adequate when producing forecasts for spare parts inventories. The need for particular spare parts may slowly increase or decrease over time (the trend component), and the trend may slowly change as different machines etc. age or become obsolete, thus affecting the trend in the demand for spare parts for the respective machines.

In order to compute the smoothed value (forecast) for the first observation in the series, both estimates of S_0 and T_0 (initial trend) are necessary. By default, these values are computed as:

$T_0 = (X_n-X_1)/(N-1)$

where N is the length of the series,

and $S_0 = X_1-T_0/2$.

Nonseasonal, no trend This Time Series model is equivalent to the simple exponential smoothing model. Note that, by default, the first smoothed value will be computed based on an initial S_0 value equal to the overall mean of the series.

Normal approximation The normal approximation to data is to approximate areas under the histogram of data, transformed into standard units, by the corresponding areas under the normal curve.

Many probability distributions can be approximated by a normal distribution, in the sense that the area under the probability histogram is close to the area under a corresponding part of the normal curve. To find the corresponding part of the normal curve, the range must be converted to standard units, by subtracting the expected value and dividing by the standard error. For example, the area under the binomial probability histogram for $n = 50$ and $p = 30\%$ between 9.5 and 17.5 is 74.2%. To use the normal approximation, we transform the endpoints to standard units, by subtracting the expected value (for the Binomial random variable, $n \times p = 15$ for these values of n and p) and dividing the result by the standard error (for a Binomial, $(n \times p \times (1-p))^{1/2} = 3.24$ for these values of n and p). The area normal approximation is the area under the normal curve between $(9.5 - 15)/3.24 = -1.697$ and $(17.5 - 15)/3.24 = 0.772$; that area is 73.5%, slightly smaller than the corresponding area under the binomial histogram. Note that the approximation gets worse when p gets close to 0 or 1, and that the approximation improves as n increases.

Normal curve The normal curve is the familiar "bell curve." The mathematical expression for the normal curve is $y = (2\times pi)^{-½}E^{-x2/2}$, where pi is the ratio of the circumference of a circle to its diameter (3.14159265 . . .), and E is the base of the natural logarithm (2.71828 . . .). The normal curve is symmetric around the point x=0, and positive for every value of x. The area under the normal curve is unity, and the SD of the normal curve, suitably defined, is also unity. Many (but not most) histograms, converted into standard units, approximately follow the normal curve.

Normal curve (distribution) A bell-shaped graph representing a hypothetical frequency distribution for a given characteristic in which most of the scores are clustered around the mean. The scores become less frequent the farther they appear above or below the mean.

Normal distribution A random variable X has a normal distribution with mean m and standard error s if for every pair of numbers $a <= b$, the chance that a < (X-m)/s < b is

P(a < (X-m)/s < b) = area under the normal curve between a and b.

If there are numbers m and s such that X has a normal distribution with mean m and standard error s, then X is said to have a normal distribution or to be normally distributed. If X has a normal distribution with mean m=0 and standard error s=1, then X is said to have a standard normal distribution. The notation $X\sim N(m,s^2)$ means that X has a normal distribution with mean m and standard error s; for example, X~N(0,1), means X has a standard normal distribution. Normal distributions model (some) continuous random variables. Strictly, a Normal random variable should be capable of assuming any value on the real line, though this requirement is often waived in practice. For example, height at a given age for a given gender in a given racial group is adequately described by a Normal random variable even though heights must be positive.

A continuous random variable X, taking all real values in the range (-∞, ∞), is said to follow a Normal distribution with parameters μ and σ, written:

$X\sim N(\mu, \sigma^2)$

If it has probability density function:

$$f(x) = \frac{1}{\sigma\sqrt{2\pi}} \exp\left[-\frac{1}{2}\left(\frac{x-\mu}{\sigma}\right)^2\right]$$

This probability density function (p.d.f.) is a symmetrical, bell-shaped curve, centred at its expected value μ. The varianceis σ^2.

Many distributions arising in practice can be approximated by a Normal distribution. Other random variables may be transformed to normality.

The simplest case of the normal distribution, known as the Standard Normal Distribution, has expected value zero and variance one. This is written as N(0,1).

Normal fit The normal/observed histogram represents the most common graphical test of normality. When you select this fit, a normal curve will be overlaid on the frequency distribution. The normal function fitted to histograms is defined as:

f(x) = NC * step * normal(x, mean, std.dev)

The normal function fitted to cumulative histograms is defined as:

f(x) = NC * inormal(x, mean, std.dev.)

where NC is the number of cases.

tep is the categorization step size

(e.g., the integral categorization step size is 1).

normal is the normal function.

inormal is the integral of the normal function.

Normal probability plots This type of graph is used to evaluate the normality of the distribution of a variable, that is, whether and to what extent the distribution of the variable follows the normal distribution. The selected variable will be plotted in a scatterplot against the values "expected from the normal distribution."

The standard normal probability plot is constructed as follows. First, the deviations from the mean (residuals) are rank ordered. From these ranks the program computes z values (i.e., standardized values of the normal distribution) based on the assumption that the data come from a normal distribution. These z values are plotted on the Y-axis in the plot. If the observed residuals (plotted on the X-axis) are normally distributed, then all values should fall onto a straight line. If the residuals are not normally distributed, then they will deviate from the line. Outliers may also become evident in this plot. If there is a general lack of fit, and the data seem to form a clear pattern (e.g., an S shape) around the line, then the variable may have to be transformed in some way .

Normality tests A common application for distribution fitting procedures is when you want to verify the assumption of normality before using some parametric test. A variety of statistics for testing normality are available including the Kolmogorov-Smirnov test for normality, the Shapiro-Wilks' W test, and the Lilliefors test. Additionally, you may review probability plots and normal probability plots to assess whether the data are accurately modeled by a normal distribution.

Normalization Adjusting a series (vector) of values (typically representing a set of measurements, e.g., a variable storing heights of people, represented in inches) according to some transformation function in order to make them comparable with some specific point of reference (for example, a unit of length or a sum). For example, dividing these values by 2.54 will produce metric measurements of the height. Normalization of data is:

(a) required when the incompatibility of the measurement units across variables may affect the results (e.g., in calculations based on cross products) without carrying any interpretable information, and

(b) recommended whenever the final reports could benefit from expressing the results in specific meaningful/compatible units (e.g., reaction time data will be easier to interpret when converted into milliseconds from the CPU cycles of different computers that were used to measure RT's - as originally registered in a medical experiment).

Note that this term is unrelated to the term normal distribution.

Not, negation, logical negation The negation of a logical proposition p, not p, is a proposition that is the logical opposite of p. That is, if p is true, not p is false, and if p is false, not p is true. Negation takes precedence over other logical operations.

Null hypothesis The null hypothesis, H0 represents a theory that has been put forward, either because it is believed to be true or because it is to be used as a basis for argument, but has not been proved. For example, in a clinical trial of a new drug, the null hypothesis might be that the new drug is no better, on average, than the current drug. We would write H0: there is no difference between the two drugs on average.

We give special consideration to the null hypothesis. This is due to the fact that the null hypothesis relates to the statement being tested, whereas the alternative hypothesis relates to the statement to be accepted if / when the null is rejected.

The final conclusion once the test has been carried out is always given in terms of the null hypothesis. We either 'reject H0 in favour of H1' or 'do not reject H0'; we never conclude 'reject H1', or even 'accept H1'.

If we conclude 'do not reject H0', this does not necessarily mean that the null hypothesis is true, it only suggests that there is not sufficient evidence against H0 in favour of H1; rejecting the null hypothesis then, suggests that the alternative hypothesis may be true.

Numerical statistics Statistical parameters presented as numbers (as opposed to pictorial statistics).

O

Object code This is the code which a computer recognises and acts upon as a direct consequence of its electromechanical construction. Typically such code is highly abstract and unsuitable for use in general use by human programmers. The object code to specify a certain process is usually generated through use of a compiler.

Objective An interpretation of probability which maintains that a probability quantifies an aspect of the real world. cf. subjective

Observational study A study in which the investigator observes but makes no intervention.

Observed counts The counts obtained from the data.

Observed frequencies In contingency table problems, the observed frequencies are the frequencies actually obtained in each cell of the table, from our random sample. When conducting a chi-squared test, the term observed frequencies is used to describe the actual data in the contingency table.

Observed frequencies are compared with the expected frequencies and differences between them suggest that the model expressed by the expected frequencies does not describe the data well.

ODBC (open database connectivity) Is a set of conventions introduced by Microsoft which allows access to information from a wide range of databases (e.g., MS Access, Oracle) and performing queries via SQL.

Odds The *odds in favor of an event* is the ratio of the probability that the event occurs to the probability that the event does not occur. For example, suppose an experiment can result in any of n possible outcomes, all equally likely, and that k of the outcomes result in a "win" and n-k result in a "loss." Then the chance of winning is k/n; the chance of not winning is $(n-k)/n$; and the odds in favor of winning are $(k/n)/((n-k)/n) = k/(n-k)$, which is the number of favorable outcomes divided by the number of unfavorable outcomes. Note that odds are not synonymous with probability, but the two can be converted back and forth. If the odds in favor of an event are q, then the probability of the event is $q/(1+q)$. If the probability of an event is p, the odds in favor of the event are $p/(1-p)$ and the odds against the event are $(1-p)/p$.

Odds ratio An alternative characterisation of the parameter 'p' for a binomial process is the ratio of the incidences of the two alternatives : p/(1-p) ; this quantity is termed the odds ratio; the value may range from zero to infinity. This relates to a possible view of a binomial process as the combined activity of two poisson processes with a limit upon total count for the two processes combined. Also see : logisitic regression.

Ogive A graphic that displays a running total.

OLE DB (object linking and embedding database) Is a set of conventions introduced by Microsoft which allows access to information from a wide range of databases (e.g., MS Access, Oracle). OLE DB is a database architecture that provides universal data integration over an enterprise's network, from mainframe to desktop, regardless of the data type. OLE DB is a more generalized and more efficient strategy for data access than ODBC, because it allows access to more types of data and is based on the component object model (COM).

On line analytic processing (OLAP) (or fast analysis of shared multidimensional information - FASMI) The term On-Line Analytic Processing refers to technology that allows users of multidimensional databases to generate on-line descriptive or comparative summaries ("views") of data and other analytic queries.

One of n encoding Representing a nominal variable using a set of input or output units, one unit for each possible nominal value. During training, one of the units will be on and the others off.

One off A case typed in and submitted to the neural network as a one-off procedure (not part of a data set, and not used in training).

One sample t test A one sample t-test is a hypothesis test for answering questions about the mean where the data are a random sample of independent observations from an underlying normal distribution:

$N(\mu, \sigma^2)$, where σ^2 is unknown

The null hypothesis for the one sample t-test is:

$H_0 : \mu = 0$ (where $\mu 0$ known)

That is, the sample has been drawn from a population of given mean and unknown variance (which therefore has to be estimated from the sample).

This null hypothesis, H0 is tested against one of the following alternative hypotheses, depending on the question posed:

$H_1 : \mu \neq \mu 0$

$H_1 : \mu > \mu 0$

$H_1 : \mu < \mu 0$

One sample test A test comparing a sample statistic with a population parameter.

One sided alternative An alternative hypothesis that asserts that a parameter value is on one side of the value(s) asserted by the null hypothesis. Such an alternative hypothesis will either say that the parameter is greater than a hypothesized value or say that it is less than a hypothesized value.

One sided confidence limit The lower or upper confidence limit when the other limit is plus or minus infinity, respectively.

One sided test A one-sided test is a statistical hypothesis test in which the values for which we can reject the null hypothesis, H0 are located entirely in one tail of the probability distribution.

In other words, the critical region for a one-sided test is the set of values less than the critical value of the test, or the set of values greater than the critical value of the test.

A one-sided test is also referred to as a one-tailed test of significance.

The choice between a one-sided and a two-sided test is determined by the purpose of the investigation or prior reasons for using a one-sided test.

Example

Suppose we wanted to test a manufacturers claim that there are, on average, 50 matches in a box. We could set up the following hypotheses

$H_0 : \mu = 50$ against $H_1 : \mu < 50$ or $H_1 : \mu > 50$

Either of these two alternative hypotheses would lead to a one-sided test. Presumably, we would want to test the null hypothesis against the first alternative hypothesis since it would be useful to know if there is likely to be less than 50 matches, on average, in a box (no one would complain if they get the correct number of matches in a box or more).

Yet another alternative hypothesis could be tested against the same null, leading this time to a two-sided test:

$H_0 : \mu = 50$ against $H_1 : \mu \neq 50$

That is, nothing specific can be said about the average number of matches in a box; only that, if we could reject the null hypothesis in our test, we would know that the average number of matches in a box is likely to be less than or greater than 50.

One sided vs. two sided limits Area outside the upper specification limit (greater than USL) is defined as one million opportunities to produce defects. Of course, in many cases any "outcomes" (e.g., parts) that are produced that fall below the specification limit can be equally defective. In that case one may want to consider the lower tail of the respective (shifted) normal distribution as well. However, in practice one usually ignores the lower tail of the normal curve because 1. In many cases, the process "naturally" has one-sided specification limits (e.g., very low delay times are not really a defect, only very long times; very few customer complaints are not a problem, only very many, etc.), and

2. When a 6 * sigma process has been achieved, the area under the normal curve below the lower specification limit is negligible.

One sided ranges or error bars in range plots In order to display a "one-sided" range (relative to the mid-point) or an error bar which extends in only one direction, set the respective values of the variable defining the range boundary to 0 (when the Relative to the Mid-point style is selected) or the mid-point (when the Absolute style is selected).

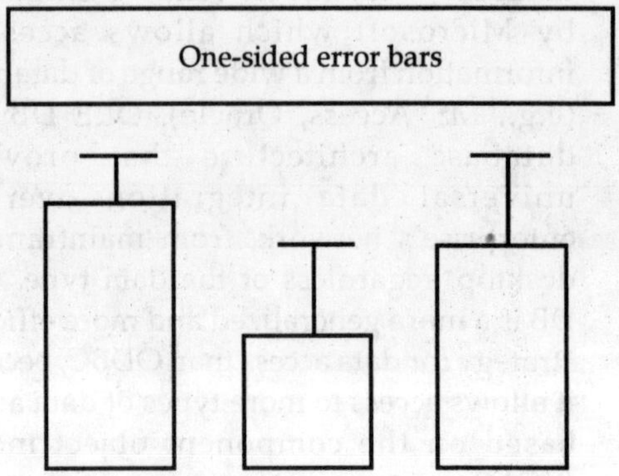

One tailed test A hypothesis test in which the rejection region consists entirely of either large values or small values of the test statistic. Such a test is appropriate against one-sided alternatives. (Note, however, that F or chi-squared tests can be one-tailed against two-sided alternatives.)

One way analysis of variance The one way analysis of variance allows us to compare several groups of observations, all of which are independent but possibly with a different mean for each group. A test of great importance is whether or not all the means are equal.

The observations all arise from one of several different groups (or have been exposed to one of several different treatments in an experiment). We are classifying 'one-way' according to the group or treatment.

Operating characteristic curves, for quality control charts A common supplementary plot to standard quality control charts is the so-called operating characteristic or OC curve. One question that comes to mind when using standard variable or attribute charts is how sensitive is the current quality control procedure? Put in more specific terms, how likely is it that you will not find a sample (e.g., a mean in an X-bar chart) outside the control limits (i.e., accept the production process as "in control"), when, in fact, it has shifted by a certain amount? This probability is usually referred to as the b (beta) error probability, that is, the probability of erroneously accepting a process (mean, mean proportion, mean rate defectives, etc.) as being "in control."

Operating characteristic curves are extremely useful for exploring the power of the quality control procedure. The actual decision concerning sample sizes should depend not only on the cost of implementing the plan (e.g., cost per item sampled), but also on the costs resulting from not detecting quality problems. The OC curve allows the engineer to estimate the probabilities of not detecting shifts of certain sizes in the production quality.

Or, disjunction, logical disjunction An operation on two logical propositions. If p and q are two propositions, (p **or**) q is a proposition that is true if p is true or if q is true (or both); otherwise, it is false. That is, (p **or**) is true unless both p and q are false.

Ordinal A scale using numbers or symbols to rank order; its intervals are unspecified.

Ordinal classification If you have a large number of classes, one-of-N encoding can become extremely unwieldy, as the number of neurons in the network proliferates. An alternative approach then is to use ordinal encoding. The output is mapped to a single neuron, with the output class represented by the ordinals 1, 2, 3, etc. The problem with this technique is that it falsely implies an ordering on the classes (i.e. class 1 is more like class 2 than class 3).

However, in some circumstances it may be the only viable approach.

You can specify ordinal encoding by changing the conversion function of the output variable to minimax. The ordinals are then mapped into the output range of the neuron.

With ordinal encoding, the output is determined as follows. The output neuron's activation is linearly scaled using the factors determined by minimax, then rounded to the nearest integer. This ordinal value gives the class.

The only classification threshold used in the accept threshold. If the difference between the output and the selected ordinal is greater than the threshold, then the classification is instead "unknown."

Example The output is 3.8, which is rounded to the nearest ordinal 4. The difference is 0.2. If an accept threshold less than 0.2 has been selected, the classification is rejected, and "unknown" is generated.

An accept threshold of 0.5 or above is equivalent to not using a threshold at all.

Ordinal data A set of data is said to be ordinal if the values / observations belonging to it can be ranked (put in order) or have a rating scale attached. You can count and order, but not measure, ordinal data.

The categories for an ordinal set of data have a natural order, for example, suppose a group of people were asked to taste varieties of biscuit and classify each biscuit on a rating scale of 1 to 5, representing strongly dislike, dislike, neutral, like, strongly like. A rating of 5 indicates more enjoyment than a rating of 4, for example, so such data are ordinal.

However, the distinction between neighbouring points on the scale is not necessarily always the same. For instance, the difference in enjoyment expressed by giving a rating of 2 rather than 1 might be much less than the difference in enjoyment expressed by giving a rating of 4 rather than 3.

Ordinal multinomial distribution If the categories for a multinomial response variable can be ordered, then the distribution of that variable is referred to as ordinal multinomial. For example, if in a survey the responses to a question are recorded such that respondents have to choose from the pre-arranged categories "Strongly agree", "Agree", "Neither agree nor disagree", "Disagree", and "Strongly disagree", then the counts (number of respondents) that endorsed the different categories would follow an ordinal multinomial distribution (since the response categories are ordered with respect to increasing degrees of disagreement).

Ordinal scale The ordinal scale of measurement represents the ranks of a variable's values. Values measured on an ordinal scale contain information about their relationship to other values only in terms of whether they are "greater than" or "less than" other values but not in terms of "how much greater" or "how much smaller."

See also, Measurement scales.

Ordinal variable A variable whose possible values have a natural order, such as {short, medium, long}, {cold, warm, hot}, or {0, 1, 2, 3, . . . }. In contrast, a variable whose possible values are {straight, curly} or {Arizona, California, Montana, New York} would not naturally be ordinal. Arithmetic with the possible values of an ordinal variable does not necessarily make sense, but it does make sense to say that one possible value is larger than another.

Ordinary least squares [OLS] The statistical procedure used in regression analysis to arrive at the 'best fitting' regression line. It is a mathematical method that ensures that the squares of the deviations from the regression line are minimised.

Outcome An outcome is the result of an experiment or other situation involving uncertainty.

The set of all possible outcomes of a probability experiment is called a sample space.

Outcome space The outcome space is the set of all possible outcomes of a given random experiment. The outcome space is often denoted by the capital letter **S.**

Outcome value The value of the test statistic for the data as initially observed, before any re-randomisation..

Outer arrays In Taguchi experimental design methodology, the repeated measurements of the response variable are often taken in a systematic fashion, with the goal to manipulate noise factors. The levels of those factors are then arranged in a so-called outer array, i.e., an (orthogonal) experimental design. However, usually the repeated measurements are placed in separate columns in the data spreadsheet (i.e., each is a different variable); thus the index i (in the formulas for smaller-the-better, larger-the-better, and signed target) runs across the columns or variables in the data spreadsheet, or the levels of the factors in the outer array.

Outer fences In a boxplot, the outer fences define the boundary beyond which lie the extreme outliers at 3.0 times the IQR above and below the upper and lower quartiles.

Outlier Outliers are atypical (by definition), infrequent observations; data points which do not appear to follow the characteristic distribution of the rest of the data. These may reflect genuine properties of the underlying phenomenon (variable), or be due to measurement errors or other anomalies which should not be modeled.

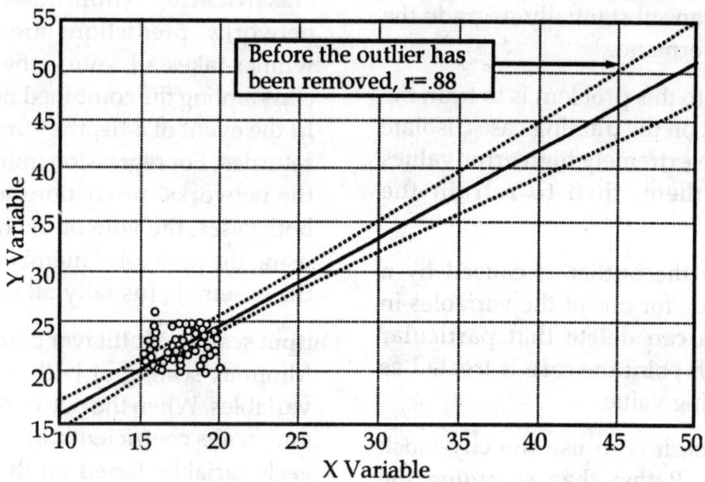

Because of the way in which the regression line is determined in Multiple Regression (especially the fact that it is based on minimizing not the sum of simple distances but the sum of squares of distances of data points from the line), outliers have a profound influence on the slope of the regression line and consequently on the value of the correlation coefficient. A single outlier is capable of considerably changing the slope of the regression line and, consequently, the value of the correlation. Note, that just one outlier can be entirely responsible for a high value of the correlation that otherwise (without the outlier) would be close to zero. Needless to say, one should never base important conclusions on the value of the correlation coefficient alone (i.e., examining the respective scatterplot is always recommended).

Note that if the sample size is relatively small, then including or excluding specific data points that are not as clearly "outliers" as the one shown in the previous example may have a profound influence on the regression line (and the correlation coefficient). This is illustrated in the following example where we call the points being excluded "outliers;" one may argue, however, that they are not outliers but rather extreme values.

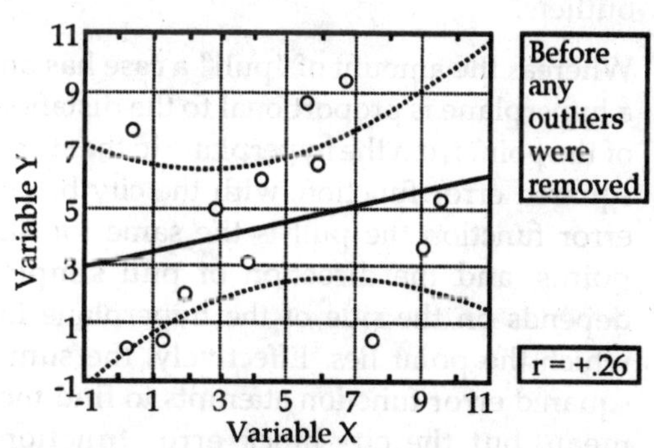

Typically, we believe that outliers represent a random error that we would like to be able to control. Needless to say, outliers may not only artificially increase the value of a correlation coefficient, but they can also decrease the value of a "legitimate"

correlation. Regression networks can be particularly prone to problems with outlying data. The use of the sum-squared network error function means that points lying far from the others have a disproportionate influence on the position of the hyperplanes used in regression. If these points are actually anomalies (for example, spurious points generated by the failure of measuring devices) they can substantially degrade the network's performance.

One approach to this problem is to train the network, test it on the training cases, isolate those that have extremely high error values and remove them, then to retrain the network.

If you believe the outlier is caused by a suspicious value for one of the variables in that case, you can delete that particular value, at which point the case is treated as having a missing value.

Another approach is to use the city-block error function. Rather than summing the squared-differences in each variable to work out an error measure, this simply sums the absolute differences. Removing the square function makes training far less sensitive to outliers.

Whereas the amount of "pull" a case has on a hyperplane is proportional to the distance of the point from the hyperplane in the sum-squared error function, with the city block error function the pull is the same for all points, and the direction of pull simply depends on the side of the hyperplane to which the point lies. Effectively, the sum-squared error function attempts to find the mean, but the city-block error function attempts to find the median.

Output ensembles Output ensembles are the most general form. Any combination of networks can be combined in an output ensemble. If the networks have different outputs, the resulting ensemble simply has multiple outputs. Thus, an output ensemble can be used to form a multiple output model where each output's prediction is formed separately.

If any networks in the ensemble have a shared output, the ensemble estimates a value for that output by combining the outputs from the individual networks. For classification (nominal outputs), the networks' predictions are combined in a winner-takes-all vote - the most common class among the combined networks is used. In the event of a tie, the "unknown" class is returned. For regression (numeric variables), the networks' predictions are averaged. In both cases, the vote or average is weighted using the networks' membership weights in the ensemble (usually all equal to 1.0).

Output scaling Multilayer perceptrons include Minimax scaling of both input and output variables. When the network is trained, shift and scale coefficients are determined for each variable, based on the minimum and maximum values in the training set, and the data is transformed by multiplying by the scale factor and adding the shift factor.

The net effect is that a 0.0 output activation level in the network is translated into the minimum value encountered in the training data, and a 1.0 activation level is translated into the maximum training data value. Consequently, the network is able to interpolate between the values represented in the training data. However, extrapolation outside the range encountered in the training set is more circumscribed. Two approaches to encoding the output are available, each of which allows a certain amount of extrapolation.

A logistic activation function is used for the output, with scaling factors determined so that the range encountered in the training set is mapped to a restricted part of the logistic functions (0,1) range (e.g. to [0.05,

0.95]. This allows a small amount of extrapolation (significant extrapolation from data is usually unjustified anyway). Using the logistic function makes training stable.

Uses an identity activation function in the final layer of the network. This supports a substantial amount of extrapolation, although not unlimited (the hidden units will saturate eventually). As a bonus, the final layer can be "fine-tuned" after iterative training using the pseudo-inverse technique. However, iterative training tends to be less stable than with a non-linear activation function, and the learning rate must be carefully chosen to avoid weight divergence during training (i.e. less than 0.1), if using an algorithm such as back propagation.

Over sampling particular strata to over represent rare events In some predictive data mining applications it is often necessary to apply stratified sampling to systematically over-sample (apply a greater sampling fraction) to particular "rare events" of interest. For example, in catalog retailing the response rate to particular catalog offers can be below 1%, and when analyzing historical data (from prior campaigns) to build a model for targeting potential customers more successfully, it is desirable to over-sample past respondents (i.e., the "rare" respondents who ordered from the catalog); one can then apply the various model building techniques for classification to a sample consisting of approximately 50% responders and 50% non-responders. Otherwise, if one were to draw a simple random sample for the analysis (with 1% of responders), then practically all model building techniques would likely predict a simple "no-response" for all cases, and would be (trivially) correct in 99% of the cases.

Overdispersion The term Overdispersion refers to the condition when the variance of an observed dependent (response) variable exceeds the nominal variance, given the respective assumed distribution. This condition occurs frequently when fitting generalized linear models to categorical response variables, and the assumed distribution is binomial, multinomial, ordinal multinomial, or Poisson. When overdispersion occurs, the standard errors of the parameter estimates and related statistics (e.g., standard errors of predicted and residual statistics) must be computed taking into account the overdispersion.

Overfitting When attempting to fit a curve to a set of data points, producing a curve with high curvature which fits the data points well, but does not model the underlying function well, its shape being distorted by the noise inherent in the data.

Overlearning When an iterative training algorithm is run, overfitting which occurs when the algorithm is run for too long (and the network is too complex for the problem or the available quantity of data).

Overparameterized model An overparameterized model uses the indicator variable approach to represent effects for categorical predictor variables in general linear models and generalized linear models. To illustrate indicator variable coding, suppose that a categorical predictor variable called Gender has two levels (i.e., Male and Female). A separate continuous predictor variable would be coded for each group identified by the categorical predictor variable. Females might be assigned a value of 1 and males a value of 0 on a first predictor variable identifying membership in the female Gender group, and males would then be assigned a value of 1 and females a value of 0 on a second predictor variable identifying membership in the male Gender group.

Note that this method of coding for categorical predictor variables will almost always lead to design matrices with redundant columns in general linear models and generalized linear models, and thus requires a generalized inverse for solving the normal equations. As such, this method is often called the overparameterized model for representing categorical predictor variables, because it results in more columns in the design matrix than are necessary for determining the relationships of the categorical predictor variables to responses on the dependent variables.

P

P (type I error) = significance level = α

The exact probability of a type II error is generally unknown.

If we do not reject the null hypothesis, it may still be false (a type II error) as the sample may not be big enough to identify the falseness of the null hypothesis (especially if the truth is very close to hypothesis).

For any given set of data, type I and type II errors are inversely related; the smaller the risk of one, the higher the risk of the other.

A type I error can also be referred to as an error of the first kind.

P (type II error) = β (but is generally unknown).

A type II error can also be referred to as an error of the second kind.

P value The probability value (p-value) of a statistical hypothesis test is the probability of getting a value of the test statistic as extreme as or more extreme than that observed by chance alone, if the null hypothesis H0, is true.

It is the probability of wrongly rejecting the null hypothesis if it is in fact true.

It is equal to the significance level of the test for which we would only just reject the null hypothesis. The p-value is compared with the significance level and, if it is smaller, the result is significant. That is, if the null hypothesis were to be rejected at

$\alpha = 0.05$, this would be reported as '$p < 0.05$'.

Small p-values suggest that the null hypothesis is unlikely to be true. The smaller it is, the more convincing is the rejection of the null hypothesis. It indicates the strength of evidence for say, rejecting the null hypothesis H0, rather than simply concluding 'reject H0' or 'do not reject H0'. Suppose we have a family of hypothesis tests of a null hypothesis that let us test the hypothesis at any significance level p between 0 and 100% we choose. The P value of the null hypothesis given the data is the smallest significance level p for which any of the tests would have rejected the null hypothesis.

For example, let X be a test statistic, and for p between 0 and 100%, let x_p be the smallest number such that, under the null hypothesis,

$P(X <= x) >= p$.

Then for any p between 0 and 100%, the rule reject the null hypothesis if $X < x_p$ tests the null hypothesis at significance level p. If we observed $X = x$, the P-value of the null hypothesis given the data would be the smallest p such that $x < x_p$.

Paired sample t test A paired sample t-test is used to determine whether there is a significant difference between the average values of the same measurement made under two different conditions. Both measurements are made on each unit in a sample, and the test is based on the paired differences between these two values. The usual null hypothesis is that the difference in the mean values is zero. For example, the yield of two strains of barley is measured in successive years in twenty different plots of agricultural land (the units) to investigate whether one crop gives a significantly greater yield than the other, on average.

The null hypothesis for the paired sample t-test is,

H0: d = u1 – u2 = 0

Where d is the mean value of the difference.

This null hypothesis is tested against one of the following alternative hypotheses, depending on the question posed.

H1: d = 0

H1: d > 0

H1: d< 0

The paired sample t-test is a more powerful alternative to a two sample procedure, such as the two sample t-test, but can only be used when we have matched samples.

Pairwise deletion of missing data vs. mean substitution In order to avoid loosing data due to casewise deletion of missing data, you can use one of two other methods. These are 1. The so-called mean substitution of missing data (replacing all missing data in a variable by the mean of that variable) and

2. Pairwise deletion of missing data. These methods of handling missing data can be requested in many modules; you can also use the mean substitution method to permanently "remove" missing data from your data set. Mean substitution offers some advantages and some disadvantages as compared to pairwise deletion. Its main advantage is that it produces "internally consistent" sets of results ("true" correlation matrices). The main disadvantages are:

Mean substitution artificially decreases the variation of scores, and this decrease in individual variables is proportional to the number of missing data (i.e., the more missing data, the more "perfectly average scores" will be artificially added to the data set).

Because it substitutes missing data with artificially created "average" data points, mean substitution may considerably change the values of correlations.

Pairwise MD deletion When pairwise deletion of missing data is selected, then cases will be excluded from any calculations involving variables for which they have missing data. In the case of correlations, the correlations between each pair of variables are calculated from all cases having valid data for those two variables.

Parameter A parameter is a value, usually unknown (and which therefore has to be estimated), used to represent a certain population characteristic. For example, the population mean is a parameter that is often used to indicate the average value of a quantity.

Within a population, a parameter is a fixed value which does not vary. Each sample drawn from the population has its own value of any statistic that is used to estimate this parameter. For example, the mean of the data in a sample is used to give information about the overall mean in the population from which that sample was drawn.

Parameters are often assigned Greek letters (e.g. σ), whereas statistics are assigned Roman letters (e.g. s).

Parametric curve Parametric equations can be used to represent curves whose graphs are not simple functions of the type y = f(x), where y and x are represented along the vertical and horizontal axes, respectively. Instead, the curves in the x-y plane are defined parametrically as two simultaneous functions of a parameter t that ranges over some interval (minimum, maximum). You can specify an equation y = f(t) for the y-component of the curve, and an equation x = g(t) for the x-component of the curve, for a specified range of parameter t.

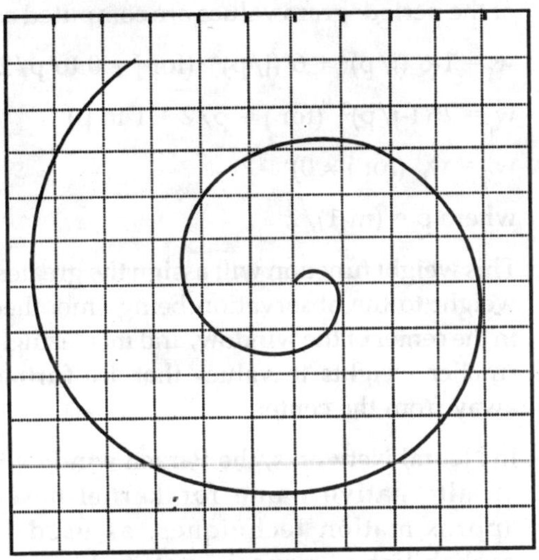

For example, to plot a spiral, you could specify:

y(t) = t*cos(t)

x(t) = t*sin(t)

For 0 < = t < = 12.

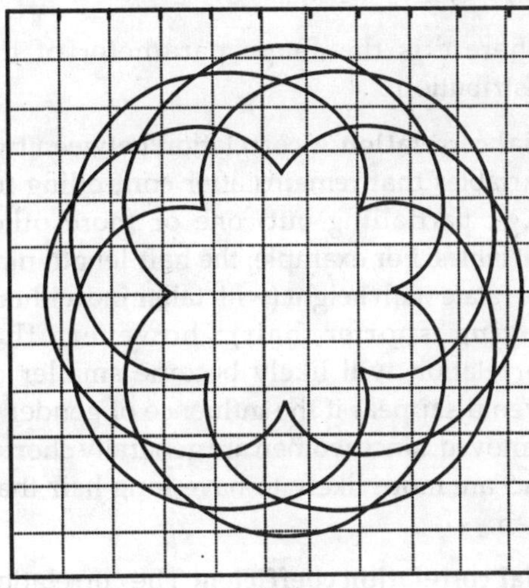

There are a wide variety of curves, from simple circles to complex shapes that can be produced via the parametric curves facilities. Here is another example:

y(t) = (a + b)*sin(t) – b*sin((a/b + 1)*t)

x(t) = (a + b)*cos(t) – b*cos((a/b + 1)* t).

Pareto chart analyses The Pareto chart analysis is a simple but powerful way of identifying the causes of quality problems or loss. It amounts to constructing a histogram of the number of quality problems or loss by some meaningful units, for example, by type of problem or division within the company.

According to the so-called Pareto principle, the majority of the quality loss is caused by a small number of factors. Put another way, in many cases, few types of problems often cause the majority of quality loss, or a few divisions within the company may produce the majority of loss. Such "maldistribution" (uneven distribution) of quality loss is very common, and can easily be detected via the Pareto chart.

Pareto distribution The standard Pareto distribution has density function (for positive parameter c):

$f(x) = c/x^{c+1} \quad 1 \le x, c > 0$

where c is the shape parameter of the distribution..

Partial correlation A correlation between two variables that remains after controlling for (e.g., partialling out) one or more other variables. For example, the hair length may correlate with height (with taller individuals having shorter hair), however, that correlation will likely become smaller or even disappear if the influence of gender is removed, since women are generally shorter and are more likely to have long hair than men.

Partial correlation coefficient The correlation coefficient between two variables when other variables have been controlled.

Partial least squares regression (PLS) Partial Least Squares is a linear regression method that forms components (factors, or latent variables) as new independent variables (explanatory variables, or predictors) in a regression model. The components in partial least squares are determined by both the response variable(s) and the predictor variables. A regression model from partial least squares can be expected to have a smaller number of components without an appreciably smaller R-square value.

Partial regression coefficient A regression coefficient describing the relationship between a dependent and an independent variables obtained when controlling for one or more other independent variables.

Partial residuals In a (linear, nonlinear, or additive) regression model with m predictors, the partial residuals for a predictor k are computed by removing from the dependent variable values the effects of all predictors i = 1, ... , m; i?k. Scatterplots of partial residuals against predictor variables are particularly useful in generalized additive models, where they may aid in the interpretation of the (non-linear) unique effect of the predictors in the overall model.

Partial table A table that displays counts for part of a dataset, for example, just for the men.

Partition A *partition* of an event B is a collection of events $\{A_1, A_2, A_3, \ldots\}$ such that the events in the collection are disjoint, and their union is B (they exhaust B). That is,

$A_j A_k = \{\}$ unless $j = k$, and

$B = A_1 \cup A_2 \cup A_3 \cup ..$

If the event B is not specified, it is assumed to be the entire outcome space s.

Parzen window In Time Series, the Parzen window is a weighted moving average transformation used to smooth the periodogram values. In the Parzen window (Parzen, 1961), for each frequency, the weights for the weighted moving average of the periodogram values are computed as:

$w_j = 1\text{-}6^*(j/p)^2 + 6^*(j/p)^3$ (for j = 0 to p/2)

$w_j = 2^*(1\text{-}j/p)^3$ (for j = p/2 + 1 to p)

$w_{-j} = w_j$ (for $j \neq 0$)

where p = (m-1)/2

This weight function will assign the greatest weight to the observation being smoothed in the center of the window, and increasingly smaller weights to values that are further away from the center.

In Neural Networks, the Parzen window is an alternative name for kernel-based approximation techniques, as used in probabilistic neural networks and generalized regression neural networks.

Pascal Named after the mathematician Blaise Pascal . A programming language designed for clarity of expression when published in human-legible form, and for the teaching of programming. Pascal is to some extent specialised for numerical work. A development is extended pascal. Compilers for pascal are widespread. Pascal and extended pascal are each represented as standard programming languages approved by ANSI and ISO.

Payoff matrix A way of representing what each player in a game wins or loses, as a function of his and his opponent's strategies.

Pearson correlation The most widely-used type of correlation coefficient is Pearson r (Pearson, 1896), also called linear or product-moment correlation (the term correlation was first used by Galton, 1888). Using non technical language, one can say that the correlation coefficient determines the extent to which values of two variables are "proportional" to each other. The value of the correlation (i.e., correlation coefficient) does not depend on the specific measurement units used; for example, the correlation between height and weight will be identical regardless of whether inches and pounds, or centimeters and kilograms are used as measurement units. Proportional means linearly related; that is, the correlation is high if it can be approximated by a straight line (sloped upwards or downwards). This line is called the regression line or least squares line, because it is determined such that the sum of the squared distances of all the data points from the line is the lowest possible. Pearson correlation assumes that the two variables are measured on at least interval scales. The Pearson product moment correlation coefficient is calculated as follows:

$r_{12} = [\Sigma (Y_{i1} - \text{Y-bar}_1)*(Y_{i2} - \text{Y-bar}_2)] / [\Sigma (Y_{i1} - \text{Y-bar}_1)^2 * \Sigma (Y_{i2} - \text{Y-bar}_2)^2]^{1/2}$

Pearson curves A system of distributions proposed by Karl Pearson consists of seven solutions (of 12 originally enumerated by Pearson) to a differential equation which approximate a wide range of distributions of different shapes. Gruska, Mirkhani, and Lamberson (1989) describe in detail how the different Pearson curves can be fit to an empirical distribution. A method for computing specific Pearson percentiles is also described in Davis and Stephens (1983).

See also, Johnson Curves.

Pearson residuals After fitting a generalized linear model to the data, to check the adequacy of the respective model, one usually computes various residual statistics. The Pearson residual is computed as the raw residual (y-m), scaled by the estimated standard deviation of y.

Pearson's product moment correlation coefficient Pearson's product moment correlation coefficient, usually denoted by r, is one example of a correlation coefficient. It is a measure of the linear association between two variables that have been measured on interval or ratio scales, such as the relationship between height in inches and weight in pounds. However, it can be misleadingly small when there is a relationship between the variables but it is a non-linear one.

There are procedures, based on r, for making inferences about the population correlation coefficient. However, these make the implicit assumption that the two variables are jointly normally distributed. When this assumption is not justified, a non-parametric measure such as the Spearman Rank Correlation Coefficient might be more appropriate.

Penalty functions A constraint specified in a loss function which applies a penalty (a very large value) to the loss function when certain undesirable conditions are met. Using a penalty function allows the user to control what permissible values of the parameters to be estimated may be manipulated by the Non-linear estimation program.

Percentage point See quantile.

Percentiles The percentile (this term was first used by Galton, 1885a) of a distribution of values is a number x_p such that a percentage p of the population values are less than or equal to x_p. For example, the 25th percentile (also referred to as the .25 quantile or lower quartile) of a variable is a value (x_p) such that 25% (p) of the values of the variable fall below that value.

Similarly, the 75th percentile (also referred to as the .75 quantile or upper quartile) is a value such that 75% of the values of the variable fall below that value and is calculated accordingly.

Perceptrons Perceptrons are a simple form of neural networks. They have no hidden layers, and can only perform linear classification tasks. Perceptrons were devised by Rosenblatt (1958), and their limitations were criticized by Minsky and Papert (1969), leading to a loss of interest in the field. Fausett (1994) gives a good history of these early developments.

A perceptron is modeled by creating a two-layer MLP network, and changing the activation function of the output layer to Step.

The perceptron learning algorithm is modeled by using back propagation with Momentum 0.0 and Shuffle turned Off.

Permutation A permutation of a set is an arrangement of the elements of the set in some order. If the set has *n* things in it, there are *n*! different orderings of its elements. For the first element in an ordering, there are *n* possible choices, for the second, there remain *n*-1 possible choices, for the third, there are *n*-2, *etc.*, and for the *n*th element of the ordering, there is a single choice remaining. By the fundamental rule of counting, the total number of sequences is thus $n \times (n-1) \times (n-2) \times \ldots \times 1$. Similarly, the number of orderings of length *k* one can form from $n >= k$ things is $n \times (n-1) \times (n-2) \times \ldots \times (n-k+1) = n!/(n-k)!$. This is denoted $_nP_k$, the number of permutations of *n* things taken *k* at a time. *C.f.* combinations.

Phi A measure of association appropriate for measuring the strength of a relationship between two dichotomous variables.

Pictorial statistics Statistical parameters that are presented as graphs or charts (as opposed to simply as numbers).

Pie chart A pie chart is a way of summarising a set of categorical data. It is a circle which is divided into segments. Each segment represents a particular category. The area of each segment is proportional to the number of cases in that category.

Example

Suppose that, in the last year a sports wear manufacturers has spent 6 million pounds on advertising their products; 3 million has been spent on television adverts, 2 million on sponsorship, 1 million on newspaper adverts, and a half million on posters.

Pie chart counts Unlike the values pie chart, this type of pie (this term was first used by

Pie chart-counts

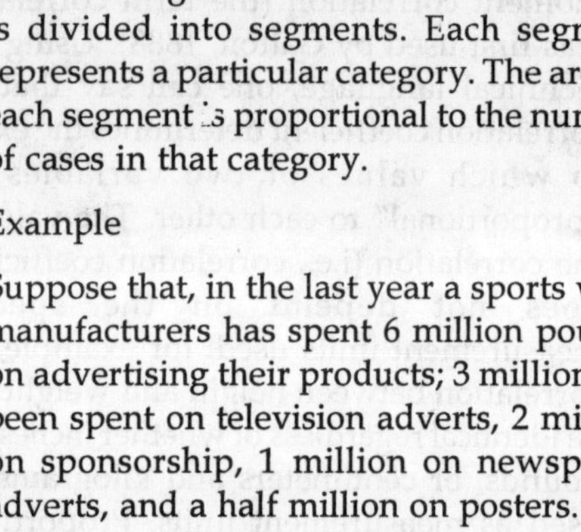

Haskell, 1922) chart (sometimes called frequency pie chart) interprets data like a histogram. It categorizes all values of the selected variable following the selected

categorization technique and then displays the relative frequencies as pie slices of proportional sizes.

Pie chart multi pattern bar The multi-pattern bar plot is designed to display the same type of data as the values pie chart (see Pie Chart - Values or 2D Histograms), however, the consecutive values are represented by the height of vertical bars (of different colors and patterns) and not areas of pie slices.

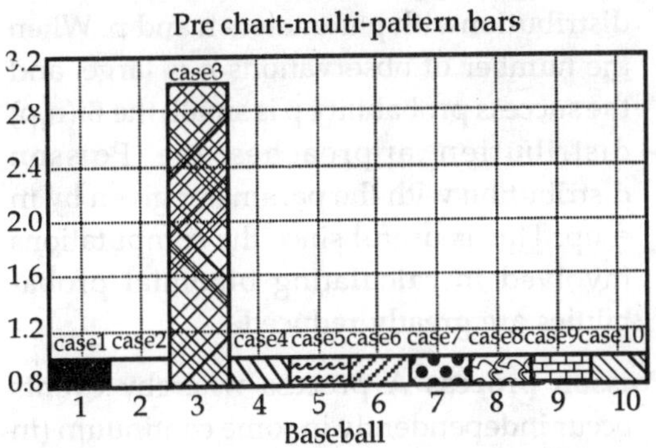

Their advantage over pie charts is that they may allow for more precise comparisons between presented values (e.g., small pie slices may be difficult to compare if they are not adjacent). This type of graph may also have advantages over regular histograms (where one fill pattern and color is used for all columns), in cases when quick identification of specific columns is desired.

See also, Pie Charts.

Pie chart values The sequence of values from the selected variable will be represented by consecutive slices of the pie (this term was first used by Haskell, 1922); the size of each slice will be proportional to the respective value. The values should be greater than 0 (0's and negative values cannot be represented as slices of the pie). This simple type of pie chart (sometimes called data pie chart) interprets data in the most straightforward manner: one case = one slice.

See also, Pie Charts.)

Pie graph A graph that represents data as percentages of a pie.

Pitman permutation test(1) Named after the statistician E.J. Pitman who described this test, and the pitman permutation test, in 1937; this is one of the earliest instances of an exact test. An exact re-randomisation test in which the test statistic is the difference of means of two samples of univariate interval-scale data.

Pitman permutation test(2) An exact re-randomisation test in which the test statistic is the mean difference of a single sample of univariate data measured under two circumstances as repeated measures.

Placebo effect The belief or knowledge that one is being treated can itself have an effect that confounds with the real effect of the treatment. Subjects given a placebo as a pain-killer report statistically significant reductions in pain in randomized experiments that compare them with subjects who receive no treatment at all. This very real psychological effect of a placebo, which has no direct biochemical effect, is called the placebo effect. Administering a placebo to the control group is thus important in experiments with human subjects; this is the essence of a blind experiment.

PNG (portable network graphics) PNG (portable network graphics) Is a graphics file format designed to store bitmap (i.e., raster, as opposed to vector/metafile) graphic images. It was introduced to be a replacement for the patented GIF format (mostly to avoid the legal restrictions associated with the patent). A PNG image may contain a variable number of colors, including a transparent color. The size of the file depends on the number of colors used in the specific image. A compression method is used (to reduce the file size) which is highly effective if a large proportion of the image is built of the pixels with the same color attributes (e.g., it is highly effective for charts and schematic line art, but somewhat less effective than the JPEG format for photographs).

Point estimate A number computed from a sample to represent a population parameter.

Point estimator An estimator that provides a single value as an estimate. cf. interval estimator.

Point of averages In a scatterplot, the point whose coordinates are the arithmetic means of the corresponding variables. For example, if the variable X is plotted on the horizontal axis and the variable Y is plotted on the vertical axis, the point of averages has coordinates (mean of X, mean of Y).

Poisson distribution Poisson distributions model (some) discrete random variables. Typically, a Poisson random variable is a count of the number of events that occur in a certain time interval or spatial area. For example, the number of cars passing a fixed point in a 5 minute interval; the number of calls received by a switchboard during a given period of time.

A discrete random variable X is said to follow a Poisson distribution with parameter m, written $X \sim Po(m)$, if it has probability distribution:

$$P(X=x) = \frac{m^x}{x!}e^{-m}$$

where:

$x = 0,1,2, ..., n$

$m > 0$.

The following requirements must be met:

a. The length of the observation period is fixed in advance;

b. The events occur at a constant average rate;

c. The number of events occurring in disjoint intervals are statistically independent.

The Poisson distribution has expected value $E(X) = m$ and variance $V(X) = m$; that is:

$$E(X) = V(X) = m$$

The Poisson distribution can sometimes be used to approximate the Binomial distributionwith parameters n and p. When the number of observations n is large, and the success probability p is small, the Bi(n,p) distribution approaches the Poisson distribution with the parameter given by m = np. This is useful since the computations involved in calculating binomial probabilities are greatly reduced.

Poisson process A process whereby events occur independently in some continuum (in many applications, time), such that the overall density (rate) is statistically constant but that it is impossible to improve any prediction of the position (time) of the next event by reference to the detail of any number of preceding observations. The corresponding distribution of intervals between events is an exponential distribution. The conventional example of a poisson processes is concerned with occurence of radioactive emissions in a substantial sample of radioactive with a half-life very much longer than the total observation period.

Polar coordinates Polar coordinates (r,θ) represent the location of a point (in 2D space) by its distance (r) from a fixed point on a fixed line (polar axis) and the angle (θ, in radians) from that fixed line.

Polar plots are used to visualize functions. They also offer an intuitive way to present relations involving a variable representing direction.

Polynomial This fits to the data, a polynomial function of the form:

$$y = b_0 + b_1x + b_2x^2 + b_3x^3 + ... + b_nx^n$$

where n is the order of the polynomial.

Pooled variance t test A t test for two samples that uses the variance estimated from all the cases in both samples. A pooled variance t test is appropriate when both samples are drawn from the same population.

Population A population is any entire collection of people, animals, plants or things from which we may collect data. It is the entire group we are interested in, which we wish to describe or draw conclusions about.

In order to make any generalisations about a population, a sample, that is meant to be representative of the population, is often studied. For each population there are many possible samples. A sample statistic gives information about a corresponding population parameter. For example, the sample mean for a set of data would give information about the overall population mean.

It is important that the investigator carefully and completely defines the population before collecting the sample, including a description of the members to be included.

Population mean The mean of the numbers in a numerical population. For example, the population mean of a box of numbered tickets is the mean of the list comprised of all the numbers on all the tickets. The population mean is a parameter. *C.f.* sample mean.

Population percentage The percentage of units in a population that possess a specified property. For example, the percentage of a given collection of registered voters who are registered as Republicans. If each unit that possesses the property is labeled with "1," and each unit that does not possess the property is labeled with "0," the population percentage is the same as the mean of that list of zeros and ones; that is, the population percentage is the population mean for a population of zeros and ones. The population percentage is a parameter. *C.f.* sample percentage.

Population standard deviation The standard deviation of the values of a variable for a population. This is a parameter, not a statistic. *C.f.* sample standard deviation.

Positive correlation The relationship between two variables is such that as one variable's values tend to increase, the other variable's values also tend to increase. This is represented by a positive correlation coefficient.

Positive predictive value For a diagnostic procedure, the conditional probability of disease given a positive test result.

Positive relationship A relationship between two variables such that when one increases, the other increases, or when one decreases, the other decreases.

Positive skew A graph that has scores bunching up toward the negative end of the abscissa.

Positively skewed curve A probability or frequency distribution that is not normal, but rather is shifted such that the mean is greater than the mode.

Positively skewed distribution A distribution in which a small proportion of individuals have excessively large values. Positively skewed distributions are characterized by having a "long right tail". For such distributions, the mean tends to be larger than the median.

Post hoc comparisons Usually, after obtaining a statistically significant F test from the ANOVA, one wants to know which means contributed to the effect; that is, which groups are particularly different from each other. One could of course perform a series of simple t-tests to compare all possible pairs of means. However, such a procedure would capitalize on chance. The reported probability levels would actually overestimate the statistical significance of mean differences. For example, suppose you took 20 samples of 10 random numbers each, and computed 20 means. Then, take the group (sample) with the highest mean and compare it with that of the lowest mean. The t-test for independent samples will test whether or not those two means are significantly different from each other, provided that they were the only two samples taken. Post-hoc comparison techniques on the other hand, specifically take into account the fact that more than two samples were taken. They are used as either hypothesis testing or exploratory methods.

Post hoc ergo propter hoc "After this, therefore because of this." A fallacy of logic known since classical times: inferring a causal relation from correlation.

Post synaptic potential (PSP) function The function applied by a unit to its inputs, weights and thresholds to form the unit's input (or activation) level. The two major PSP functions are linear (weighted sum minus threshold) and radial (scaled squared distance of weight vector from input vector).

Posterior probability The probability assigned to a statement after making an observation. cf. prior probability. Such probabilities are typically epistemic and subjective.

Potential capability (C_p) This is the simplest and most straightforward indicator of process capability. It is defined as the ratio of the specification range to the process range; using ± 3 sigma limits we can express this index as:

C_p = (USL-LSL)/(6*Sigma)

Put into words, this ratio expresses the proportion of the range of the normal curve that falls within the engineering specification limits (provided that the mean is on target, that is, that the process is centered).

Power The power of a statistical hypothesis test measures the test's ability to reject the null hypothesis when it is actually false - that is, to make a correct decision.

In other words, the power of a hypothesis test is the probability of not committing a type II error. It is calculated by subtracting the probability of a type II error from 1, usually expressed as:

Power = 1 – P(type II error) = (1 – β)

The maximum power a test can have is 1, the minimum is 0. Ideally we want a test to have high power, close to 1. This is the probability that a statistical test will detect a defined pattern in data and declare the extent of the pattern as showing statistical significance. Power is related to type-2 error by the simple formula : power = (1-beta) ; the motive for this re-definition is so that an increase in value for power shall represent improvement of performance of a statistical test.

Power (statistical) See statistical power.

Power goal The minimum power to be achieved when searching for an acceptable sample size. An acceptable sample size must yield a power greater than or equal to this value.

Pre coded A question in a survey questionnaire is pre-coded when a number has been assigned to every possible response in advance of the survey.

Precise estimator An estimator with a small standard error.

Precision Precision is a measure of how close an estimator is expected to be to the true value of a parameter.

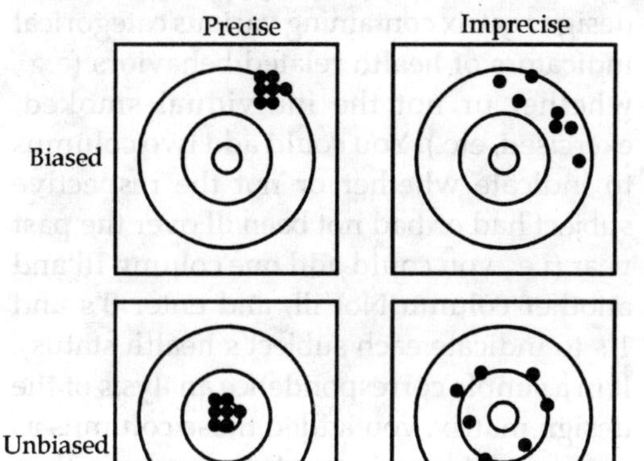

Precision is usually expressed in terms of imprecision and related to the standard error of the estimator. Less precision is reflected by a larger standard error.

The following illustrates bias and precision, where the target value is the bullseye:

Precise Imprecise

Biased Unbiased

Example

The police decide to estimate the average speed of drivers using the fast lane of the motorway and consider how it can be done. One method suggested is to tail cars using police patrol cars and record their speeds as being the same as that of the police car. This is likely to produce a biased result as any driver exceeding the speed limit will slow down on seeing a police car behind them. The police then decide to use an unmarked car for their investigation using a speed gun operated by a constable. This is an unbiased method of measuring speed, but is imprecise compared to using a calibrated speedometer to take the measurement.

Predicted value A value of a response variable predicted by a statistical model. cf. fitted value.

Prediction On discovering an association or relationship between two or more variables in a sample, a prediction may be made about how other cases in the population, that were not measured, may behave under the same conditions.

Prediction interval An interval that includes a specified proportion of individuals in a population having a particular value (or values) of an explanatory variable. The confidence level associated with a prediction interval is the probability that it will include a randomly chosen individual having the specified characteristics. cf. confidence interval.

Prediction interval ellipse Often plotted on 2D Scatterplots, this interval describes the area in which a single new observation can be expected to fall with a certain probability (alpha), given that the new observation comes from a bivariate normal distribution with the parameters (means, standard deviations, covariance) as estimated from the observed points shown in the plot.

The coordinates for the ellipse are computed so that:

$((n-p)*n)/(p*(n-1)*(n+1))*(X-X_m)'S^{-1}(X-X_m)\sim F(alpha,p,n-p)$

where: n number of cases

p number of variables; i.e., p=2 in the case of the bivariate scatterplot

X vector of coordinates (pair of coordinates, since p=2)

X_m vector of means for the p dimensions (variables) in the plot

S^{-1} inverse of the variance covariance matrix for the p variables

F(alpha,p, n-p) the value of F, given alpha, p, and n-p.

Note that if the number of observations in the scatterplot is small, then the prediction interval may be very large.

Prediction profiles When the results of an experiment are analyzed, the observed responses on the dependent variables can be fit to a separate prediction equation for each dependent variable (containing different coefficients but the same terms). Once these equations are constructed, predicted values for the dependent variables can be computed at any combination of levels of the predictor variables. A prediction profile for a dependent variable consists of a series of graphs, one for each independent variable, of the predicted values for the dependent variable at different levels of one independent variable, holding the levels of the other independent variables constant at specified values. Inspecting the prediction profiles for the dependent variables can show which levels of the predictor variables produce the most desirable predicted responses on the dependent variables.

Predictive data mining The term Predictive Data Mining is usually applied to identify data mining projects with the goal to identify a statistical or neural network model or set of models that can be used to predict some response of interest. For example, a credit card company may want to engage in predictive data mining, to derive a (trained) model or set of models (e.g., neural networks, meta-learner) that can quickly identify transactions which have a high probability of being fraudulent. Other types of data mining projects may be more exploratory in nature (e.g., to identify cluster or segments of customers), in which case drill-down descriptive and exploratory methods would be applied. Data reduction is another possible objective for data mining (e.g., to aggregate or amalgamate the information in very large data sets into useful and manageable chunks).

Predictive mapping One application of multiple correspondence analysis is to perform the equivalent of a Multiple Regression for categorical variables, by adding supplementary columns to a design matrix. For example, suppose you had a design matrix containing various categorical indicators of health related behaviors (e.g., whether or not the individual smoked, exercised, etc.). You could add two columns to indicate whether or not the respective subject had or had not been ill over the past year (i.e., you could add one column Ill and another column Not ill, and enter 0's and 1's to indicate each subject's health status). If in a simple correspondence analysis of the design matrix, you added those columns as supplementary columns to the analysis, then

1. The summary statistics for the quality of representation for those columns would give you an indication of how well you can "explain" illness as a function of the other variables in the design matrix, and

2. The display of the column points in the final coordinate system would provide an indication of the nature (e.g., direction) of the relationships between the columns in the design matrix and the column points indicating illness. This technique (adding supplementary points to a multiple correspondence analysis) is also called predictive mapping.

Predictive model markup language (PMML) PMML (short for Predictive Model Markup Language) is an XML-based language which allows for the efficient exchange of (trained) predictive models and shared models between different applications. A PMML document usually contains information describing fully trained or parameterized analytic models so that they can be readily deployed (applied to new cases) by another application.

Predictors Predictors (also called independent or input variables) are variables used to predict or explain the value(s) of one or more dependent variables (also referred to as dependent or outcome variables).

Press statistic The press statistic is often used in regression analyses, in order to summarize the fit of a particular model in a sample of observations that were not used to estimate the model parameters. It can simply be computed as the sums of squares of the prediction residuals for those observations.

The press statistic is, for example, computed in the Partial Least Squares Model for the cross-validation (verification) samples.

Prevalence More properly called point prevalence rate.

Of a disease, the proportion of a population having the disease at a given point in time.

Prima facie Latin for "at first glance." "On the face of it." *Prima facie* evidence for something is information that at first glance supports the conclusion. On closer examination, that might not be true; there could be another explanation for the evidence.

Primary analysis Analysis of data collected by the researcher. See secondary analysis

Principal components analysis A linear dimensionality reduction technique, which identifies orthogonal directions of maximum variance in the original data, and projects the data into a lower-dimensionality space formed of a sub-set of the highest-variance components.

See also, Factor Analysis.

Prior probabilities Proportionate distribution of classes in the population (in a classification problem), especially where known to be different than the distribution in the training data set. Used to modify probabilistic neural network training in neural networks.

See Discriminant Function Analysis.

Probabilistic neural networks (PNN) A type of neural network using kernel-based approximation to form an estimate of the probability density functions of classes in a classification problem. One of the so-called Bayesian networks.

Probability A probability provides a quantatative description of the likely occurrence of a particular event. Probability is conventionally expressed on a scale from 0 to 1; a rare event has a probability close to 0, a very common event has a probability close to 1.

The probability of an event has been defined as its long-run relative frequency. It has also been thought of as a personal degree of belief that a particular event will occur (subjective probability).

In some experiments, all outcomes are equally likely. For example if you were to choose one winner in a raffle from a hat, all raffle ticket holders are equally likely to win, that is, they have the same probability of their ticket being chosen. This is the equally-likely outcomes model and is defined to be:

$$P(E) = \frac{\text{number of outcomes corresponding to event E}}{\text{total number of outcomes}}$$

Examples

1. The probability of drawing a spade from a pack of 52 well-shuffled playing cards is:

$$\frac{13}{52} = \frac{1}{4} = 0.25$$

since:

Event E = 'a spade is drawn'

The number of outcomes corresponding to E = 13 (spades) and The total number of outcomes = 52 (cards).

2. When tossing a coin, we assume that the results 'heads' or 'tails' each have equal probabilities of 0.5.

Probability density function The chance that a continuous random variable is in any range of values can be calculated as the area under a curve over that range of values. The curve is the probability density function of the random variable. That is, if X is a continuous random variable, there is a function *f*(x) such that for every pair of numbers *a*<=*b*,

P(*a*<= X <=*b*) = (area under *f* between *a* and *b*);

f is the probability density function of X. For example, the probability density function of a random variable with a standard normal distribution is the normal curve. Only continuous random variables have probability density functions. The probability density function of a continuous random variable is a function which can be integrated to obtain the probability that the random variable takes a value in a given interval.

More formally, the probability density function, f(x), of a continuous random variable X is the derivative of the cumulative distribution function F(x):

$$f(x) = \frac{d}{dx} F(x)$$

Since $F(x) = P(X \le x)$ it follows that:

$$\int f(x)\, dx = F(b) - F(a) = P(a < X < b)$$

If f(x) is a probability density function then it must obey two conditions:

a. That the total probability for all possible values of the continuous random variable X is 1:

$$\int f(x)\, dx = 1$$

b. That the probability density function can never be negative: f(x) > 0 for all x.

Probability distribution The probability distribution of a random variable specifies the chance that the variable takes a value in any subset of the real numbers. (The subsets have to satisfy some technical conditions that are not important for this course.) The probability distribution of a random variable is completely characterized by the cumulative probability distribution function; the terms sometimes are used synonymously. The probability distribution of a discrete random variable can be characterized by the chance that the random variable takes each of its possible values. For example, the probability distribution of the total number of spots S showing on the roll of two fair dice can be written as a table:

s	P(S=s)
2	1/36
3	2/36
4	3/36
5	4/36
6	5/36
7	6/36
8	5/36
9	4/36
10	3/36
11	2/36
12	1/36

The probability distribution of a continuous random variable can be characterized by its probability density function.

Probability histogram A probability histogram for a random variable is analogous to a histogram of data, but instead of plotting the area of the bins proportional to the relative frequency of observations in the class interval, one plots the area of the bins proportional to the probability that the random variable is in the class interval.

Probability law Also called probability distribution.

A rule assigning probabilities to events. The term distribution is usually used to refer to probabilities associated with a random variable.

Probability mass function (abbreviated pmf) also called probability function, discrete probability density function, probability distribution function.

A function associated with a random variable that assigns to each possible value the probability that the random variable takes that value.

Probability probability (P P) plot A probability-probability (P-P) plot is used to see if a given set of data follows some specified distribution. It should be approximately linear if the specified distribution is the correct model.

The probability-probability (P-P) plot is constructed using the theoretical cumulative distribution function, F(x), of the specified model. The values in the sample of data, in order from smallest to largest, are denoted x(1),x(2),.....,x(n).

For i = 1, 2, ..., n

F(x(i)) is plotted against: $\frac{i-\frac{1}{2}}{n}$. You can visually check for the fit of a theoretical distribution to the observed data by examining the probability-probability plot (also called Probability Plot). In probability-probability plots (or P-P plots for short) the observed cumulative distribution function is plotted against the theoretical cumulative distribution function. As in the Quantile-Quantile plot, the values of the respective variable are first sorted into ascending order. The ith observation is plotted against one axis as i/n (i.e., the observed cumulative distribution function), and against the other axis as $F(x_{(i)})$, where $F(x_{(i)})$ stands for the value of the theoretical cumulative distribution function for the respective observation $x_{(i)}$. If the theoretical cumulative distribution approximates the observed distribution well, then all points in this plot should fall onto the diagonal line.

Probability probability plots categorized In this graph, you can visually check for the fit of a theoretical distribution to the observed data by examining each probability-probability plot (also called Probability Plot, see also Probability-Probability Plots) for the respective level of the grouping variable (or user-defined subset of data). In probability-probability plots (or P-P plots for short) the observed cumulative distribution function is plotted against the theoretical cumulative distribution function. As in the Categorized Quantile-Quantile plot, the values of the respective variable are first sorted into ascending order. The ith observation is plotted against one axis as i/n (i.e., the observed cumulative distribution function), and against the other axis as $F(x_{(i)})$, where $F(x_{(i)})$ stands for the value of the theoretical cumulative distribution function for the respective observation $x_{(i)}$. If the theoretical cumulative distribution approximates the observed distribution well, then all points in this plot should fall onto the diagonal line. One component graph is produced for each level of the grouping variable (or user-defined subset of data) and all the component graphs are arranged in one display to allow for comparisons between the subsets of data (categories).

Probability sample A sample drawn from a population using a random mechanism so that every element of the population has a known chance of ending up in the sample.

Probability sampling In probability sampling, every observation in the population from which the sample is drawn has a known probability of being selected into the sample; when that probability is the same for every observation in the population, the sample is an equal probability sample or epsem sample (equal probability of selection method).

Epsem samples have certain desirable properties; for example, the simple formulas for computing means, standard deviations, and so on can be applied to estimate the respective parameters in the population.

Probability, theories of A *theory of probability* is a way of assigning meaning to probability statements such as "the chance that a thumbtack lands point-up is 2/3." That is, a theory of probability connects the mathematics of probability, which is the set of consequences of the axioms of probability, with the real world of observation and experiment. There are several common theories of probability. According to the *frequency theory of probability*, the probability of an event is the limit of the percentage of times that the event occurs in repeated, independent trials under essentially the same circumstances. According to the *subjective theory of probability*, a probability is a number that measures how strongly we believe an event will occur. The number is on a scale of 0% to 100%, with 0% indicating that we are completely sure it won't occur, and 100% indicating that we are completely sure that it will occur. According to the theory of *equally likely outcomes*, if an experiment has *n* possible outcomes, and (for example, by symmetry) there is no reason that any of the *n* possible outcomes should occur preferentially to any of the others, then the chance of each outcome is 100%/*n*. Each of these theories has its limitations, its proponents, and its detractors.

Probit regression and transformation In the probit regression model, the predicted values for the dependent variable will never be less than (or equal to) 0, or greater than (or equal to) 1, regardless of the values of the independent variables; it is, therefore, commonly used to analyze binary dependent or response variables (see also the binomial distribution). This is accomplished by applying the following regression equation (the term probit was first used by Bliss, 1934):

$y = NP(b_0 + b_1 * X_1 ...)$

where NP stands for normal probability (space under the normal distribution; or cumulative distribution function of the normal distribution). One can easily recognize that, regardless of the regression coefficients or the magnitude of the x values, this model will always produce predicted values (predicted y's) in the range of 0 to 1.

Process analysis In industrial settings, Process Analysis refers to a collection of analytic methods which can be used to ensure adherence of a product to quality specifications. These methods include Sampling Plans, Process (Machine) Capability Analysis, fitting measurements to Non-Normal Distributions, analysis of Gage Repeatability and Reproducibility and Weibull and Reliability/Failure Time Analysis.

Process capability indices In industrial quality control, once a process is in control, indices are often computed to measure the quality of the items produced (and thus the capability of the process); specifically, the extent to which the items that are produced fall within allowable engineering tolerances. Given a sample of a particular size, one can estimate the standard deviation of the respective quality characteristic of interest (e.g., piston ring diameters); one can then draw a histogram of the distribution of the characteristic of interest (piston ring diameters).

If the distribution of the quality characteristic is normal, one can make inferences concerning the proportion of items (piston rings) within specification limits (methods for non-normal distributions are also available; namely the so-called Percentile Method).

Common indices for process capability are C_p, C_r, and C_{pk}. All of these indices basically reflect the ratio of the engineering tolerances (process range) to the ± 3 sigma limits around the nominal specifications.

Process performance indices When monitoring a process via a quality control chart (e.g., the X-bar and R-chart; Quality Control) it is often useful to compute the process capability indices for the process. Specifically, when the data set consists of multiple samples, such as data collected for the quality control chart, then one can compute two different indices of variability in the data. One is the regular standard deviation for all observations, ignoring the fact that the data consist of multiple samples; the other is to estimate the process's inherent variation from the within-sample variability. For example, when plotting X-bar and R-charts one may use the common estimator R-bar/d2 for the process sigma. Note however, that this estimator is only valid if the process is statistically stable.

When the total process variability is used in the standard capability computations, the resulting indices are usually referred to as process performance indices (as they describe the actual performance of the process; common indices are P_p, P_r, and P_{pk}), while indices computed from the inherent variation (within-sample sigma) are referred to as capability indices (since they describe the inherent capability of the process; common indices are C_p, C_r, and C_{pk}).

Product limit estimator See kaplan-Meier estimator

Program A sequence of instructions expressed in some programming language.

Programmable The characteristic of a computer which enables it to be used to undertake a variety of different processes on different occasions.

Programming language A formal code for expressing to a computer how a certain process should be undertaken. The translation from the code of the programming language to the object code of the appropriate computer is itself undertaken by a program for that computer; the translation program may take the form of either a compiler of an interpreter.

Proportion For a binomial random event, the probability of a successful (or favorable) outcome in a single trial.

Proportional hazards A model for survivor functions in which the corresponding hazards maintain a constant ratio over time. In a typical situation, there are a number of groups of individuals, with all the individuals in a group assumed to have the same survivor function. In this situation, proportional hazards is equivalent to the idea that there is a fixed odds-ratio for death in one group over the other that does not depend on the time of that death.

Proportional reduction in error The increased accuracy (i.e. reduction in error) in predicting the characteristics of sample on one variable that one obtains if the values of a second variable are known, compared with not knowing the second variable. For example, three-quarters of all those in work, work full-time. Knowing only this, one can predict that there is a chance of 3 in 4 that any person in the sample is a full-timer. The chances of error in the predication are reduced if one also knows that a person is female, because we know that, for women, 55 per cent are in full-time work. The proportional reduction of error is related to the association between the two variables.

Proposition, logical proposition A logical proposition is a statement that can be either true or false. For example, "the sun is shining in Berkeley right now" is a proposition.

Prospective An adjective describing an observational study in which information on individuals is collected forward in time. (Also called a *cohort* study.)

Pruning (in classification trees) Pruning a classification tree refers to the use of the automatic "right-sized" tree selection procedures developed by Breiman et. al. (1984). These procedures are designed to find, without relying on subjective judgment, the "right-sized" classification tree, that is, a classification tree with an appropriate number of splits and optimal predictive accuracy.

Pseudo components Pseudo-components are transformed values of the components (plotted in Ternary graphs) where:

$x'_i = (x_i - L_i)/(\text{Total}-L)$

Here, x'_i stands for the ith pseudo-component, x_i stands for the original component value, L_i stands for the lower constraint (limit) for the ith component, L stands for the sum of all lower constraints (limits) for all components in the design, and Total stands for the mixture total. This transformation makes the coefficients for different factors comparable in size.

Pseudo inverse algorithm An algorithm to efficiently optimize a linear model; also known as singular value decomposition.

Pseudo inverse singular value decomposition This algorithm uses the singular value decomposition technique to calculate the pseudo-inverse of the matrix needed to set the weights in a linear (dot product synaptic function + identity activation function) output layer, so as to find the least mean squared solution. Essentially, it guarantees to find the optimal setting for the weights in a linear layer, to minimize the RMS training set error (Bishop, 1995; Press et. al., 1992; Golub and Kahan, 1965). This is the standard least-squares optimization technique.

Linear techniques are extremely important in optimization, not least because it is possible to find an optimal solution to a linear model - something that is not guaranteed with nonlinear models, such as other types of neural networks, even if training algorithms converge.

The pseudo-inverse procedure, in addition to guaranteeing to find the absolute minimum error, is also relatively quick.

Pseudo-inverse is typically used in a number of circumstances:

To optimize the linear output layer in a radial basis function network, subsequent to center and deviation assignment using the unsupervised algorithms.

To optimize the output layer in a linear network.

To fine-tune the final layer in a multilayer perceptron with a linear output layer, as used in regression problems.

Pseudo random A source of data which is effectively unpredictable although generated by a determinate process. Successive pseudo-random data are produced by a fixed calculation process acting upon preceding data from the pseudo-random sequence. To start the sequence it is necessary to decide arbitrarily upon a first datum, which is termed the seed value. Also see : bootstrap, Monte-Carlo test.

Pseudo random number A number generated from a complex formula designed for the purpose, usually by computer, which has the properties of a random number

Pure error For certain designs with replicates at the levels of the predictor variables, the residual sum of squares can be further partitioned into meaningful parts which are relevant for testing hypotheses. Specifically, the residual sums of squares can be partitioned into lack-of-fit and pure-error components. This involves determining the part of the residual sum of squares that can be predicted by including additional terms for the predictor variables in the model (for example, higher-order polynomial or interaction terms), and the part of the residual sum of squares that cannot be predicted by any additional terms (i.e., the sum of squares for pure error). A test of lack-of-fit can then be performed, using the mean square pure error as the error term.

Q

Quadratic A function which fits a second-order polynomial function to the points in the 3D scatterplot.

Qualitative variable Phenomenon measured in kind, that is, non-numerical units. For example, color is a qualitative variable, because it cannot be expressed simply as a number.

Quality The term Quality in the context of correspondence analysis pertains to the quality of representation of the respective row point in the coordinate system defined by the respective numbers of dimensions, as chosen by the user. The quality of a point is defined as the ratio of the squared distance of the point from the origin in the chosen number of dimensions, over the squared distance from the origin in the space defined by the maximum number of dimensions (remember that the metric in the typical correspondence analysis is Chi-square). By analogy to Factor Analysis, the quality of a point is similar in its interpretation to the communality for a variable in factor analysis.

A low quality means that the current number of dimensions does not represent well the respective row (or column).

Quality control In all production processes, the extent to which products meet quality specifications must be monitored. In the most general terms, there are two "enemies" of product quality: 1. Deviations from target specifications, and

2. Excessive variability around target specifications. During the earlier stages of developing the production process, designed experiments are often used to optimize these two quality characteristics (see Experimental Design).

The general approach to on-line quality control is straightforward: We simply extract samples of a certain size from the ongoing production process. We then produce line charts of the variability in those samples, and consider their closeness to target specifications. If a trend emerges in those lines, or if samples fall outside pre-specified limits, then we declare the process to be out of control and take action to find the cause of the problem. These types of charts are sometimes also referred to as Shewhart control charts (named after W. A. Shewhart who is generally credited as being the first to introduce these methods).

Quantile (of a variable) The qth quantile is a value such that a fraction q of the individuals do not exceed that value, and a fraction 1-q have at least that value. (cf. :percentile) The

1/4, 1/2, and 3/4 quantiles are generally called lower quartile, median, and upper quartile, respectively.

(of a distribution) The qth (lower) quantile is a value such that the probability of being less than that value is no more than q and the probabiity of being greater than it is no more than 1-q. The qth upper quantile is the same as the 1-qth lower quantile. Quantiles are also (loosely) called perentage points, particularly when used in hypothesis testing.

Quantile quantile (QQ) plot A quantile-quantile (Q-Q) plot is used to see if a given set of data follows some specified distribution. It should be approximately linear if the specified distribution is the correct model.

The quantile-quantile (Q-Q) plot is constructed using the theoretical cumulative distribution function, F(x), of the specified model. The values in the sample of data, in order from smallest to largest, are denoted x(1), x(2),....., x(n).

For i = 1, 2,....., n,

x(i) is plotted against: $F^{-1}\left(\frac{i - \frac{1}{2}}{n}\right)$

You can visually check for the fit of a theoretical distribution to the observed data by examining the quantile-quantile (or Q-Q) plot (also called Quantile Plot).

In this plot, the observed values of a variable are plotted against the theoretical quantiles. A good fit of the theoretical distribution to the observed values would be indicated by this plot if the plotted values fall onto a straight line. To produce a Q-Q plot, the program will first sort the n observed data points into ascending order, so that:

$x_1 \leq x_2 \leq ... \leq x_n$

These observed values are plotted against one axis of the graph; on the other axis the plot will show:

$F^{-1}((i-r_{adj}) / (n+n_{adj}))$

Where i is the rank of the respective observation, r_{adj} and n_{adj} are adjustment factors (≤ 0.5) and F^{-1} denotes the inverse of the probability integral for the respective standardized distribution. The resulting plot is a scatterplot of the observed values against the (standardized) expected values, given the respective distribution. Note also that the adjustment factors r_{adj} and n_{adj} ensure that the p-value for the inverse probability integral will fall between 0 and 1, but not including 0 and 1.

Quantile quantile plots categorized In this graph, you can visually check for the fit of a theoretical distribution to the observed data by examining each quantile-quantile (or Q-Q) plot (also called Quantile Plot, see Quantile-Quantile Plots) for the respective level of the grouping variable (or user-defined subset of data).

In this plot, the observed values of a variable are plotted against the theoretical quantiles. A good fit of the theoretical distribution to the observed values would be indicated by this plot if the plotted values fall onto a straight line. One component graph is produced for each level of the grouping variable(or user-defined subset of data) and all the component graphs are arranged in one display to allow for comparisons between the subsets of data (categories).

Quantitative variable A variable that takes numerical values for which arithmetic makes sense, for example, counts, temperatures, weights, amounts of money, *etc.* For some variables that take numerical values, arithmetic with those values does not make sense; such variables are not quantitative. For example, adding and subtracting social security numbers does not make sense. Quantitative variables typically have units of measurement, such as inches, people, or pounds.

Quartile Quartiles are values that divide a sample of data into four groups containing (as far as possible) equal numbers of observations.

A data set has three quartiles. There are three quartiles. The first or lower quartile (LQ) of a list is a number (not necessarily a number in the list) such that at least 1/4 of the numbers in the list are no larger than it, and at least 3/4 of the numbers in the list are no smaller than it. The second quartile is the median. The third or upper quartile (UQ) is a number such that at least 3/4 of the entries in the list are no larger than it, and at least 1/4 of the numbers in the list are no smaller than it. To find the quartiles, first sort the list into increasing order. Find the smallest integer that is at least as big as the number of entries in the list divided by four. Call that integer *k*. The *k*th element of the sorted list is the lower quartile. Find the smallest integer that is at least as big as the number of entries in the list divided by two. Call that integer *l*. The *l*th element of the sorted list is the median. Find the smallest integer that is at least as large as the number of entries in the list times 3/4. Call that integer *m*. The *m*th element of the sorted list is the upper quartile.

References to quartiles often relate to just the outer two, the upper and the lower quartiles; the second quartile being equal to the median. The lower quartile is the data value a quarter way up through the ordered data set; the upper quartile is the data value a quarter way down through the ordered data set.

Example

Data 6 47 49 15 43 41 7 39 43 41 36.

Ordered Data6 7 15 36 39 41 41 43 43 47.

Median 41.

Upper quartile 43.

Lower quartile 15.

Quartile range The quartile (this term was first used by Galton, 1882) range of a variable is calculated as the value of the 75th percentile minus the value of the 25th percentile. Thus it is the width of the range about the median that includes 50% of the cases.

See Nonparametrics

Quasi Newton method Quasi-Newton is an advanced method of training multilayer perceptrons. It usually performs significantly better than Back Propagation, and can be used wherever back propagation can be. It is the recommended technique for most networks with a small number of weights (less than a couple of hundred). If the network is a single output regression network and the problem has low residuals, then Levenberg-Marquardt may perform better.

Quasi-Newton is a batch update algorithm: whereas back propagation adjusts the network weights after each case, Quasi-Newton works out the average gradient of the error surface across all cases before updating the weights once at the end of the epoch.

For this reason, there is no shuffle option available with Quasi-Newton, since it would clearly serve no useful function. There is also no need to select learning or momentum rates for Quasi-Newton, so it can be much easier to use than back propagation. Additive noise would destroy the assumptions made by Quasi-Newton about the shape of search space, and so is also not available.

Quasi-Newton works by exploiting the observation that, on a quadratic (i.e. parabolic) error surface, one can step directly to the minimum using the Newton step - a calculation involving the Hessian matrix (the matrix of second partial derivatives of the error surface). Any error surface is approximately quadratic "close to" a minimum. Since, unfortunately, the Hessian matrix is difficult and expensive to calculate, and anyway the Newton step is likely to be wrong on a non-quadratic surface, Quasi-Newton iteratively builds up an approximation to the inverse Hessian. The approximation at first follows the line of steepest descent, and later follows the estimated Hessian more closely.

Quasi-Newton is the most popular algorithm in nonlinear optimization, with a reputation for fast convergence. It does, however, has some drawbacks - it is rather less numerically stable than, say, Conjugate Gradient Descent, it may be inclined to converge to local minima, and the memory requirements are proportional to the square of the number of weights in the network.

It is often beneficial to precede Quasi-Newton training with a short burst of Back Propagation (say 100 epochs), to cut down on problems with local minima.

If the network has many weights, you are advised to use Conjugate Gradient Descent instead. Conjugate Gradient Descent has memory requirements proportional only to the number of weights, not the square of the number of weights, and the training time is usually comparable with Quasi-Newton, if somewhat slower.

Quest Quest is a classification tree program developed by Loh and Shih (1997).

Quick propagation Despite the name, quick propagation (Fahlman, 1988; Patterson, 1996) is not necessarily faster than back propagation, although it may prove significantly faster for some applications.

Quick propagation also sometimes seems more inclined to instability and to getting stuck in local minima, than back propagation; these tendencies may determine whether quick propagation is more appropriate for a particular problem.

Quick propagation is a batch update algorithm: whereas back propagation adjusts the network weights after each case, quick propagation works out the average gradient of the error surface across all cases before updating the weights once at the end of the epoch.

For this reason, there is no shuffle option available with quick propagation, since it would clearly serve no useful function.

Quick propagation works by making the (typically ill-founded) assumption that the error surface is locally quadratic, with the axes of the hyper-ellipsoid error surface aligned with the weights. If this is true, then the minimum of the error surface can be found after only a couple of epochs. Of course, the assumption is not generally valid, but if it is even close to true, the algorithm can converge on the minimum very rapidly.

Based on this assumption, quick propagation works as follows:

On the first epoch, the weights are adjusted using the same rule as back propagation, based upon the local gradient and the learning rate.

On subsequent epochs, the quadratic assumption is used to attempt to move directly to the minimum.

The basic quick propagation formula suffers from a number of numerical problems. First, if the error surface is not concave, the algorithm can actually go the wrong way. If the gradient changes little or not at all, the change can be extremely large, or even infinite! Finally, if a zero gradient is encountered, a weight will stop changing permanently.

Quintile Quintiles are values that divide a sample of data into five groups containing (as far as possible) equal numbers of observations.

See quantile.

Compare quartile, percentile.

Quota sampling Quota sampling is a method of sampling widely used in opinion polling and market research. Interviewers are each given a quota of subjects of specified type to attempt to recruit for example, an interviewer might be told to go out and select 20 adult men and 20 adult women, 10 teenage girls and 10 teenage boys so that they could interview them about their television viewing.

It suffers from a number of methodological flaws, the most basic of which is that the sample is not a random sample and therefore the sampling distributions of any statistics are unknown. Quota sampling usually refers to the process whereby a researcher attempts to match in a sample the exact makeup of the population with regard to certain demographic characteristics deemed important (such as gender, age, race, income, etc.). For example, a researcher may strive to draw a sample from a population so that the sample consists of exactly 50% males and 50% females, certain percentages of persons from particular ethnic backgrounds, etc. The purpose of this practice usually is to achieve some kind of representative sample of the underlying population.

In general, only properly drawn probability samples such as EPSEM samples will guarantee that the population to which one wishes to generalize is properly "represented."

R

R square A measure that describes how well a regression line fits the scatter of data points. It describes the proportion of variance of the dependent variable that is 'explained' by the independent variable. In simple regression, it is the square of the correlation coefficient, r. Also known as the coefficient of determination. The ratio of regression sum of squares to residual sum of squares, representing the proportion of variability explained by the model. R^2 is typical expressed as a percent.

Radial basis functions A type of neural network employing a hidden layer of radial units and an output layer of linear units, and characterized by reasonably fast training and reasonably compact networks. Introduced by Broomhead and Lowe (1988) and Moody and Darkin (1989), they are described in most good neural network text books.

Radial sampling Radial sampling is a simple technique to assign centers to radial units in the first hidden layer of a network by randomly sampling training cases and copying those to the centers. This is a reasonable approach if the training data are distributed in a representative manner for the problem (Lowe, 1989).

The number of training cases must at least equal the number of centers to be assigned.

Ramp A piece-wise linear version of the sigmoid function. Relatively poor training performance, but fast execution.

Random An event for which there is no way to know, before it occurs, what the outcome will be. Instead, only the probabilities of each possible outcome can be stated.

Random effects (in mixed model ANOVA) The term random effects in the context of analysis of variance is used to denote factors in an ANOVA design with levels that were not deliberately arranged by the experimenter (those factors are called fixed effects), but which were sampled from a population of possible samples instead. For example, if one were interested in the effect that the quality of different schools has on academic proficiency, one could select a sample of schools to estimate the amount of variance in academic proficiency (component of variance) that is attributable to differences between schools.

A simple criterion for deciding whether or not an effect in an experiment is random or fixed is to ask how one would select (or arrange) the levels for the respective factor in a replication of the study. For example, if one wanted to replicate the study described in this example, one would choose (take a sample of) different schools from the population of schools. Thus, the factor "school" in this study would be a random factor. In contrast, if one wanted to compare the academic performance of boys to girls in an experiment with a fixed factor Gender, one would always arrange two groups: boys and girls. Hence, in this case the same (and in this case only) levels of the factor Gender would be chosen when one wanted to replicate the study.

Random error All measurements are subject to error, which can often be broken down into two components: a bias or systematic error, which affects all measurements the same way; and a random error, which is in general different each time a measurement is made, and behaves like a number drawn with replacement from a box of numbered tickets whose average is zero.

Random event See random experiment.

Random experiment An experiment or trial whose outcome is not perfectly predictable, but for which the long-run relative frequency of outcomes of different types in repeated trials is predictable. Note that "random" is different from "haphazard," which does not necessarily imply long-term regularity.

Random number seed The starting value for a generator of pseudo-random numbers. The stream of pseudo-random numbers from the generator is always the same for the same starting seed.

Random numbers from arbitrary distributions Random numbers can be generated for all continuous and discrete distributions, using the standard inversion method. First generate a uniform random number, and then use the inverse distribution function for the distribution of interest to generate the random variate values. In practice, care must be taken not to use uniform random numbers that are (almost) 0 or (almost) 1, since the inverse distribution functions at the extreme margins may return in some cases the missing data code. For example, suppose you want to generate random variate values for the Weibull distribution with scale parameter .5, shape parameter .6, and location (threshold) parameter 10. You could use a formula to generate those values:

```
=invWeibull(rnd(1)*.99999+.000001,.5,.6,10)
```

Note that the rnd(1) function will generate uniform random numbers in the range from 0 to 1; by multiplying those values by .99999 and adding .000001 you guarantee that the inverse Weibull distribution function vWeibull is not called with a very small (almost 0) or very large (almost 1) probability value (where it might return a missing value; the constants used in this example guarantee .000001<=p<=.999991); thus the variable computed in this manner is guaranteed to contain only valid random values from the specified Weibull distribution (except for the extreme tails).

Random sample A random sample is a sample whose members are chosen at random from a given population in such a way that the chance of obtaining any particular sample can be computed. The number of units in the sample is called the *sample size*, often denoted n. The number of units in the population often is denoted N. Random samples can be drawn with or without replacing objects between draws; that is, drawing all n objects in the sample at once (a random sample without replacement), or drawing the objects one at a time, replacing them in the population between draws (a random sample with replacement). In a random sample with replacement, any given member of the population can occur in the sample more than once. In a random sample without replacement, any given member of the population can be in the sample at most once. A random sample without replacement in which every subset of n of the N units in the population is equally likely is also called a simple random sample. The term *random sample with replacement* denotes a random sample drawn in such a way that every n-tuple of units in the population is equally likely. See also probability sample.

Random sampling Random sampling is a sampling technique where we select a group of subjects (a sample) for study from a larger group (a population). Each individual is chosen entirely by chance and each member of the population has a known, but possibly non-equal, chance of being included in the sample.

By using random sampling, the likelihood of bias is reduced.

Compare simple random sampling.

Random sub sampling in data mining When mining huge data sets with many millions of observations, it is neither practical nor desirable to process all cases (although efficient incremental learning algorithms exist to perform predictive data mining using all observations in the dataset). For example, by properly sampling only 100 observations (from millions of observations) you can compute a very reliable estimate of the mean. One of the rules of statistical sampling that is often not intuitively understood by untrained "observers" is the fact that the reliability and validity of results depend, among many other things, on the size of a random sample, and not on the size of the population from which it is taken. In other words, the mean estimated from 100 randomly sampled observations is as accurate (i.e., falls within the same confidence limits) regardless of whether the sample was taken from 1000 cases or 100 billion cases. Put another way, given a certain (reasonable) degree of accuracy required, there is absolutely no need to process and include all observations in the final computations (for estimating the mean, fitting models, etc.).

Random variable A random variable is an assignment of numbers to possible outcomes of a random experiment. For example, consider tossing three coins. The number of heads showing when the coins land is a random variable: it assigns the number 0 to the outcome {T, T, T}, the number 1 to the outcome {T, T, H}, the number 2 to the outcome {T, H, H}, and the number 3 to the outcome {H, H, H}. The outcome of an experiment need not be a number, for example, the outcome when a coin is tossed can be 'heads' or 'tails'. However, we often want to represent outcomes as numbers. A random variable is a function that associates a unique numerical value with every outcome of an experiment. The value of the random variable will vary from trial to trial as the experiment is repeated.

There are two types of random variable - discrete and continuous.

Examples

1. A coin is tossed ten times. The random variable X is the number of tails that are noted. X can only take the values 0,1,....,10, so X is a discrete random variable.

2. A light bulb is burned until it burns out. The random variable Y is its lifetime in hours. Y can take any positive real value, so Y is a continuous random variable.

A random variable has either an associated probability distribution (discrete random variable) or probability density function (continuous random variable).

Randomisation(1) Generation of whole or part of the randomisation set. Also See randomisation(3), re-randomisation.

Randomisation(2) The process of arranging for data-collection, in accordance with the experimental design, such that there should be no foreseeable possibilty of any systematic relationship between the data and any measureable characteristic of the procedure by which the data was sampled. This is usually arranged by assigning experimental units to groups, and repeated measures to experimental units, on a strictly random basis.

Randomisation(3) One of the arrangements making up the randomisation set. These arranegments will be encountered in the act of randomisation(1). Also See branch and bound, minimal-change sequence.

Randomisation distribution A collection of values of the test statistic obtained by undertaking a number of re-randomisations of the actual data within the randomisation set. ALso see : confidence interval, randomisation test.

Randomisation set The collection of possible re-randomisations of data within the constraints of the experimental design. Also see randomisation distribution.

Randomisation test The rationale of a randomisation test involves exploring re-randomisations of the actual data to form the randomisation distribution of values of the test statistic. The outcome value value of the test statistic is judged in terms of its relative position within the re-randomisation distribution. If the outcome value is near to one extreme of the re-randomisation distribution then it may be judged that it is in the extreme tail of the distribution, with reference to a nominal alpha criterion value, and thus judged to show statistical significance. Also See exact test.

Randomised complete block design The randomised complete block design is a design in which the subjects are matched according to a variable which the experimenter wishes to control. The subjects are put into groups (blocks) of the same size as the number of treatments. The members of each block are then randomly assigned to different treatment groups.

Example

A researcher is carrying out a study of the effectiveness of four different skin creams for the treatment of a certain skin disease. He has eighty subjects and plans to divide them into 4 treatment groups of twenty subjects each. Using a randomised blocks design, the subjects are assessed and put in blocks of four according to how severe their skin condition is; the four most severe cases are the first block, the next four most severe cases are the second block, and so on to the twentieth block. The four members of each block are then randomly assigned, one to each of the four treatment groups.

Randomized controlled experiment An experiment in which chance is deliberately introduced in assigning subjects to the treatment and control groups. For example, we could write an identifying number for each subject on a slip of paper, stir up the slips of paper, and draw slips without replacement until we have drawn half of them. The subjects identified on the slips drawn could then be assigned to treatment, and the rest to control. Randomizing the assignment tends to decrease confounding of the treatment effect with other factors, by making the treatment and control groups roughly comparable in all respects but the treatment.

Range The range of a sample (or a data set) is a measure of the spread or the dispersion of the observations. It is the difference between the largest and the smallest observed value of some quantitative characteristic and is very easy to calculate.

A great deal of information is ignored when computing the range since only the largest and the smallest data values are considered; the remaining data are ignored.

The range value of a data set is greatly influenced by the presence of just one unusually large or small value in the sample (outlier).

Examples

1. The range of 65,73,89,56,73,52,47 is 89-47 = 42.

2. If the highest score in a 1st year statistics exam was 98 and the lowest 48, then the range would be 98-48 = 50.

Range ellipse This type of ellipse is a fixed size ellipse determined such that the length of its horizontal and vertical projection onto the X- and Y-axis (respectively) is equal to the mean (Range * I) where the mean and range refer to the X or Y variable, and I is the current value of the coefficient field.

Range plots boxes In this style of range plot, the range is represented by a "box" (i.e., as a rectangular box where the top of the box is the upper range and the bottom of the box is the lower range). The midpoints are represented either as point markers or horizontal lines that "cut" the box.

Range plots columns In this style of range plot, a column represents the mid-point (i.e., the top of the column is at the mid- point value) and the range (represented by "whiskers") is overlaid in the column.

Range plots whiskers In this style of range plot, the range is represented by "whiskers" (i.e., as a line with a serif on both ends). The midpoints are represented by point markers.

Rank A rank is a consecutive number assigned to a specific observation in a sample of observations sorted by their values, and thus reflecting the ordinal relation of the observation to others in the sample. Depending on the order of sorting (ascending or descending), the higher ranks represent the higher values (i.e., ascending ranks, the lowest value is assigned a rank of 1, and the highest value - the "last" (highest) rank) or higher ranks represent the lower values (i.e., descending ranks, the highest value is assigned a rank of 1).

Rank correlation A rank correlation coefficient is a coefficient of correlation between two random variables that is based on the ranks of the measurements and not the actual values, for example.

Ranked data This refers to the practice of taking a set of n data, to be regarded as ordinal-scale, amd replacing each datum by its rank (1 .. N) within the set.

Ratio A scale using numbers to rank order; its intervals are equal, and the scale has an absolute 0 point.

Ratio scale This is a type of measurement scale for which it is meaningful to reason in terms of differences in scores (see interval scale) and also in terms of ratios of scores. Such a scale will have a zero point which is meaningful in the sense that it indicates complete absence of the property which the scale measures. The ratio scale may be either unipolar (negative values not meaningful) or bipolar (both positive and negative values meaningful), and either continuous or discrete.

Ratio variable A ratio variable is one that can be measured on an interval scale and also has a meaningful zero. For instance, income is a ratio variable because it is possible to have a zero income. See level of measurement.

Rayleigh distribution The Rayleigh distribution has the probability density function:

$f(x) = x/b^2 * e^{-(x^2/2b^2)}$

$0 \leq x < \infty$

$b > 0$

where b is the scale parameter

e is the base of the natural logarithm, sometimes called Euler's e

See also, Process Analysis.

Re randomisation statistics Also known as permutation or randomisation(1) statistics.

Re randomisation The process of generating alternative arrangements of given data which would be consistent with the experimental design. Also See **bootstrap, exact test(2), exhaustive re-randomisation, re-randomisation statistics.**

Real class intervals When grouping data into more convenient categories, it is usual to declare real class intervals that are one level of precision greater than the original data. This ensures that every possible value can be accounted for in the new coding scheme. For example, if age data have been collected to the nearest whole year, the real class intervals would be reported to one decimal place. For example, the stated interval of 10 to 19 years would be expressed as a real class interval of 9.5 to 19.5.

Real class limits Real class limits are the upper and lower values that contain a real class interval.

Receiver operating characteristic (ROC) curve When a neural network is used for classification, confidence levels (the Accept and Reject thresholds, available from the pre/ post processing editor) determine how the neural networks assigns input cases to classes.

In the case of two-class classification problems, by default the output class is indicated by a single output neuron, with high output corresponding to one class and low output to the other. If the Reject threshold is strictly less than the Accept threshold, then the network may include a "doubt" option, where it is not sure of the class if the output lies between the Reject and Accept thresholds.

An alternative approach is to set the Accept and Reject thresholds equal. In this case, as the single decision threshold is adjusted, so the classification behavior of the network changes. At one extreme all cases will be assigned to one class, and at the other extreme to the other. In between these extremes, different compromises may be found, leading to different trade-offs between the rate of erroneous assignment to each class (i.e. false-positives and false-negatives).

A Receiver Operating Characteristic curve (Zweig, 1993) summarizes the performance of a two-class classifier across the range of possible thresholds. It plots the sensitivity (class two true positives) versus one minus the specificity (class one false negatives). An ideal classifier hugs the left side and top side of the graph, and the area under the curve is 1.0. A random classifier should achieve approximately 0.5 (a classifier with an area less than 0.5 can be improved simply by flipping the class assignment). The ROC curve is recommended for comparing classifiers, as it does not merely summarize performance at a single arbitrarily selected decision threshold, but across all possible decision thresholds.

The ROC curve can be used to select an optimum decision threshold. This threshold (which equalizes the probability of misclassification of either class; i.e. the probability of false-positives and false-negatives) can be used to automatically set confidence thresholds in classification networks with a nominal output variable with the Two-state conversion function.

Recoding Recoding in spss is the procedure used to change or regroup the numeric codes of a variable. For example, if age is coded using respondents' ages in years and you wish to group individuals into those above and below the age of 40, you could recode age so that codes from 0 through 40 are changed to code 1 and all other codes are changed to code 2.

Rectangular distribution The rectangular distribution (continuous uniform distribution) is useful for describing random variables with a constant probability density over the defined range $a<b$:

$f(x) = 1/(b-a) \quad a<x<b$

$\quad\;\; = 0$ elsewhere

where

$a<b$ are constants

Region of acceptance The area of a probability curve in which a computed test statistic will lead to acceptance of the null hypothesis.

Region of rejection The area of a probability curve in which a computed test statistic will lead to rejection of the null hypothesis.

Regression A category of problems where the objective is to estimate the value of a continuous output variable from some input variables.

See also multiple regression.

Regression In regression problems the purpose is to predict the value of a continuous output variable. Regression problems can done in Neural Networks using multilayer perceptrons, radial basis function networks, (Bayesian) regression networks, and linear networks.

Regression coefficient Part of the regression equation which describes the extent of the relationship between the dependent and an independent variable. In the regression equation Y=a +bX , a, the constant, is the intercept on the Y axis and b is a measure of the slope. The standardised b coefficient is called the beta coefficient.

Regression equation A regression equation allows us to express the relationship between two (or more) variables algebraically. It indicates the nature of the relationship between two (or more) variables. In particular, it indicates the extent to which you can predict some variables by knowing others, or the extent to which some are associated with others.

A linear regression equation is usually written

$Y = a + bX + e$

where

Y is the dependent variable.

a is the intercept.

b is the slope or regression coefficient.

X is the independent variable.

e is the error term.

The equation will specify the average magnitude of the expected change in Y given a change in X.

The regression equation is often represented on a scatterplot by a regression line.

Regression fallacy The regression fallacy is to attribute the regression effect to an external cause.

Regression line A regression line is a line drawn through the points on a scatterplot to summarise the relationship between the variables being studied. When it slopes down (from top left to bottom right), this indicates a negative or inverse relationship between the variables; when it slopes up (from bottom right to top left), a positive or direct relationship is indicated.

The regression line often represents the regression equation on a scatterplot.

Regression plane The analogue of a regression line when there are two independent variables. The plane plots in three dimensional space the expected values of the dependent variable for all values of the two independent variables.

Regression sum of squares The sum of squared deviations of fitted values from the overall mean, representing the amount of variability explained by the model.

Regression summary statistics In regression problems, the purpose of the neural network is to learn a mapping from the input variables to a continuous output variable, or variables.

A network is successful at regression if it makes predictions that are more accurate than a simple estimate.

The simplest way to construct an estimate, given training data, is to calculate the mean of the training data, and use that mean as the predicted value for all previously unseen cases.

The average expected error from this procedure is the standard deviation of the training data. The aim in using a regression network is therefore to produce an estimate that has a lower prediction error standard deviation than the training data standard deviation.

Regression toward the mean, regression effect Suppose one measures two variables for each member of a group of individuals, and that the correlation coefficient of the variables is positive (negative). If the value of the first variable for that individual is above average, the value of the second variable for that individual is likely to be above (below) average, but by fewer standard deviations than the first variable is. That is, the second observation is likely to be closer to the mean in standard units. For example, suppose one measures the heights of fathers and sons. Each individual is a (father, son) pair; the two variables measured are the height of the father and the height of the son. These two variables will tend to have a positive correlation coefficient: fathers who are taller than average tend to have sons who are taller than average. Consider a (father, son) pair chosen at random from this group. Suppose the father's height is 3SD above the average of all the fathers' heights. (The SD is the standard deviation of the fathers' heights.) Then the son's height is also likely to be above the average of the sons' heights, but by fewer than 3SD (here the SD is the standard deviation of the sons' heights).

In an hypothesis test using a test statistic, the rejection region is the set of values of the test statistic for which we reject the null hypothesis.

Regression, linear regression Linear regression fits a line to a scatterplot in such a way as to minimize the sum of the squares of the residuals. The resulting regression line, together with the standard deviations of the two variables or their correlation coefficient, can be a reasonable summary of a scatterplot if the scatterplot is roughly football-shaped. In other cases, it is a poor summary. If we are regressing the variable Y on the variable X, and if Y is plotted on the vertical axis and X is plotted on the horizontal axis, the regression line passes through the point of averages, and has slope equal to the correlation coefficient times the SD of Y divided by the SD of X.

Regular histogram This simple histogram will produce a column plot of the frequency distribution for the selected variable (if more than one variable is selected, then one graph will be produced for each variable in the list).

Regularization A modification to training algorithms which attempts to prevent over- or under-fitting of training data by building in a penalty factor for network complexity (typically by penalizing large weights, which correspond to networks modeling functions of high curvature).

Rejection region A set of numbers such that the null hypothesis is rejected precisely in the event that the value of the test statistic falls inside the set. Typically, rejection regions are defined as all the numbers greater than (or equal to) a critical value.

Relative frequency Relative frequency is another term for proportion; it is the value calculated by dividing the number of times an event occurs by the total number of times an experiment is carried out. The probability of an event can be thought of as its long-run relative frequency when the experiment is carried out many times.

If an experiment is repeated n times, and event E occurs r times, then the relative frequency of the event E is defined to be:

$$rf_n(E) = \frac{r}{n}$$

Example

Experiment Tossing a fair coin 50 times n = 50

Event E = 'heads'.

Result 30 heads, 20 tails r = 30.

Relative frequency:

$$rfn(E) = \frac{r}{n} = \frac{30}{50} = \frac{3}{5} = 0.6$$

If an experiment is repeated many, many times without changing the experimental conditions, the relative frequency of any particular event will settle down to some value. The probability of the event can be defined as the limiting value of the relative frequency:

$$P(E) = \lim_{n \to \infty} rfn(E)$$

For example, in the above experiment, the relative frequency of the event 'heads' will settle down to a value of approximately 0.5 if the experiment is repeated many more times.

Relative frequency principle of probability If a random event is repeated a large number of times, then the proportion of times that a particular outcome occurs is the probability of that outcome occurring in a single event.

Relative function change criterion The relative function change criterion is used to stop iteration when the function value is no longer changing (see Structural Equation Modeling). Basically, it stops iteration when the function ceases to change. The criterion is necessary because, sometimes, it is not possible to reduce the discrepancy function even when the gradient is not close to zero. This occurs, in particular, when one of the parameter estimates is at a boundary value. The "true minimum," where the gradient actually is zero, includes parameter values that are not permitted (like negative variances, or correlations greater than one).

On the i'th iteration, this criterion is equal to

$$FC_i = \left| \frac{F_i - F_{i-1}}{1 + |F_i|} \right|$$

Relative power A comparison of two or more statistical tests, for the same experimental design, sample size, and nominal alpha criterion value, in terms of the respective values of power. Also see : beta.

Relative risk (also called risk ratio.) The probability of occurrence of a disease in the presence of a putative causal agent divided by the probability of occurrence in the absence of the agent.

Reliability and item analysis In many areas of research, the precise measurement of hypothesized processes or variables (theoretical constructs) poses a challenge by itself. For example, in psychology, the precise measurement of personality variables or attitudes is usually a necessary first step before any theories of personality or attitudes can be considered. In general, in all social sciences, unreliable measurements of people's beliefs or intentions will obviously hamper efforts to predict their behavior. The issue of precision of measurement will also come up in applied research, whenever variables are difficult to observe. For example, reliable measurement of employee performance is usually a difficult task; yet, it is obviously a necessary precursor to any performance-based compensation system.

In all of these cases, Reliability & Item Analysis may be used to construct reliable measurement scales, to improve existing scales, and to evaluate the reliability of scales already in use. Specifically, Reliability & Item Analysis will aid in the design and evaluation of sum scales, that is, scales that are made up of multiple individual measurements (e.g., different items, repeated measurements, different measurement devices, etc.). Reliability & Item Analysis provides numerous statistics that allow the user to build and evaluate scales following the so-called classical testing theory model.

The term reliability used in industrial statistics denotes a function describing the probability of failure (as a function of time). In this context reliability is defined as the extent to which a measurement taken with multiple-item scale (e.g., questionnaire) reflects mostly the so-called true score of the dimension that is to be measured, relative to the error. A similar notion of scale reliability is sometimes used when assessing the accuracy (and reliability) of gages or scales used in quality control charting.

Reliability Reliability is concerned with whether the indicator we use to measure a concept gives the same answer each time it is used. See validity.

Repeated measures This is a feature of an experimental design whereby several observations measured on a common scale refer to the same sampling unit. Identification of the relation of the individual observations to the experimental design is crucial to this definition. Examples : the measurement of water level at a particular site on several systematically-defined occasions; measurement of reaction-time of an individual using right hand and left hand separately. Also See independent groups, replications, stratified.

Replications This is a feature of an experimental design whereby observations on an experimental unit are repeated under the same conditions. Identification of the position of a particular observation within the sequence of replications is irrelevant. Also See repeated measures, stratified.

Representative Patterns in a sample of units may reasonably be attributed to the population from which the sample is drawn, only if the sample is representative. In practical terms, to ensure that a sample is representative almost always means ensuring that it is a random sample.

Representative sample The notion of a "representative sample" is often misunderstood. The general intent usually is to draw a sample from a population so that particular properties of that population can be estimated accurately from the sample. For example, political scientists may draw samples from the population of voters to predict with some certainty the outcome of an election.

In general, only properly drawn probability samples such as EPSEM samples will guarantee that the population to which one wishes to generalize is properly "represented." On the other hand, a generally erroneous notion is commonly expressed that, in order to achieve "representativeness," it is desirable to draw a stratified sample using particular "quotas" (quota sampling) where demographic characteristics such as age, gender, race, and so on are properly "balanced," to match precisely the makeup of the underlying population. This notion is false: The precision of the estimates (such as voting margins) for a population computed from such a sample will only be enhanced, if the variables that one is attempting to match (age, gender, race...) are (strongly) related to the outcome variable of interest (e.g., voting behavior). However, in practice such a-priori knowledge is usually elusive, and applying such quota sampling methods may yield grossly misleading results.

Resampling A major problem with neural networks is the generalization issue (the tendency to overfit the training data), accompanied by the difficulty in quantifying likely performance on new data.

This difficulty can be disturbing if you are accustomed to the relative security of linear modeling, where a given set of data generates a single "optimal" linear model. However, this security may be somewhat deceptive, and if the underlying function is not linear, the model may be very far from optimal.

In contrast, in nonlinear modeling some choice must be made about the complexity (curvature, eccentricity) of the model, and this can lead to a plethora of alternative models. Given this diversity, it is important to have ways to estimate the performance of the models on new data, and to be able to select among them.

Most work on assessing performance in neural modeling concentrates on approaches to resampling. A neural network is optimized using a training subset. Often, a separate subset (the selection subset) is used to halt training to mitigate over-learning, or to select from a number of models trained with different parameters. Then, a third subset (the test subset) is used to perform an unbiased estimation of the network's likely performance.

Although the use of a test set allows us to generate unbiased performance estimates, these estimates may exhibit high variance. Ideally, we would like to repeat the training procedure a number of different times, each time using new training, selection and test cases drawn from the population - then, we could average the performance prediction over the different test subsets, to get a more reliable indicator of generalization performance.

In reality, we seldom have enough data to perform a number of training runs with entirely separate training, selection and test subsets. However, intuitively we might think we can do better if we train multiple networks, as when a single network is trained, only part of the data is actually involved in training. Can we find a way to use all the data in training, selection and test?

Cross validation is the most simple resampling technique. Suppose that we decide to conduct ten experiments with a given data set. We divide the data set into ten equal parts. Then, for each experiment we select one part to act as the test set. The other nine tenths of the data set are used for training and selection. When the ten experiments are finished, we can average the test set performances of the individual networks.

Cross validation has some obvious advantages. If training a single network, we would probably reserve 25% of the data for test. By using cross validation, we can reduce the individual test set size. In the most extreme version, leave-one-out cross validation, we perform a number of experiments equal to the size of the data set. On each experiment a single case is placed in the test subset, and the rest of the data is used for training. Clearly this may require a substantial number of experiments if the data set is large, but it can give you a very accurate estimate of generalization performance.

What precisely does cross validation tell us? In cross validation, each of the set of experiments should be performed with the same process parameters (same training algorithms, number of epochs, learning rates, etc.). The averaged performance measure is then an estimate of the performance on new data (drawn from the same distribution as the training data) of a single network trained using the same procedure (including the networks actually generated in the cross validation procedure).

We could select one of the cross validated networks at random and deploy it, using the estimates generated in cross validation to characterize its expected performance. However, this seems intuitively wasteful - having generated a number of networks, why not use them all? We can form the networks into an ensemble, and make predictions by averaging or voting across the resampled member networks (ensembles can also usefully combine the predictions of networks trained using different parameters, or of different architectures).

If we form an ensemble from the cross validated networks, is the performance estimate formed by averaging the test set performance of the individual networks an unbiased estimate of generalization performance?

The answer is: no. The expected performance of an ensemble is not, in general, the same as the average performance of the members. Actually, the expected performance of the ensemble is at least the average performance of the members, but usually better. Thus you can use the estimate so-formed, knowing that it is conservatively biased.

Cross validation is one technique for resampling data. There are others:

Random (Monte Carlo) resampling - the subsets are randomly sampled from the available cases. Each available case is assigned to one of the three subsets.

Bootstrapping - this technique (Efron, 1979) samples a data set with replacement (i.e. a single case may be randomly sampled several times into the bootstrap set). The bootstrap can be applied any number of times, for increased accuracy. Compared with random sampling, the use of sampling with replacement can help to iron out generalization problems caused by the finite size of the data set. Breiman (1996) suggested using the bootstrap sampling technique to train multiple models for ensemble averaging (in his case the models were decision trees, but the conclusions carry over to other models), a technique he refers to as bagging.

Resampling stats This is the name of an educational initiative involving the use of a programming language, in the form of an interpreter, allowing the user to specify monte-carlo resampling of a set of data and accumulation of the randomisation distribution of a defined test statistic.

Research hypothesis A prediction or expectation to be tested. If the null hypothesis is rejected, then the research hypothesis (also called alternative hypothesis) is supported.

Residual Residuals are differences between the observed values and the corresponding values that are predicted by the model and thus they represent the variance that is not explained by the model. The better the fit of the model, the smaller the values of residuals. The ith residual (e_i) is equal to:

$e_i = (y_i - y_i\text{-hat})$

where y_i is the ith observed value

y_i-hat is the corresponding predicted value. The difference between a datum and the value predicted for it by a model. In linear regression of a variable plotted on the vertical axis onto a variable plotted on the horizontal axis, a residual is the "vertical" distance from a datum to the line. Residuals can be positive (if the datum is above the line) or negative (if the datum is below the line). Plots of residuals can reveal computational errors in linear regression, as well as conditions under which linear regression is inappropriate, such as nonlinearity and heteroscedasticity. If linear regression is performed properly, the sum of the residuals from the regression line must be zero; otherwise, there is a computational error somewhere.

Residual plot A residual plot for a regression is a plot of the residuals from the regression against the explanatory variable.

Residual sum of squares (also called error sum of squares). The sum of squares of the residuals, representing the amount of variation not explained by the model.

Resistant A statistic is said to be resistant if corrupting a datum cannot change the statistic much. The mean is not resistant; the median is. See also breakdown point.

Resistant measure A statistic that is less likely to be affected by a few extreme high or low values (outliers) in a distribution. The median is a resistant measure which is not affected by extreme values. The mean is not a resistant measure.

Resolution An experimental design of resolution R is one in which no l-way interactions are confounded with any other interaction of order less than R - l. For example, in a design of resolution R equal to 5, no l = 2-way interactions are confounded with any other interaction of order less than R - l = 3, so main effects are unconfounded with each other, main effects are unconfounded with 2-way interactions, and 2-way interactions are unconfounded with each other.

Response surface A surface plotted in three dimensions, indicating the response of one or more variable (or a neural network) as two input variables are adjusted with the others held constant.

Response variable Also called predictand or dependent variable.

A variable whose values are to be explained by other variables. In scatterplots the response variable is usually plotted on the vertical axis.

Retrospective Adjective describing an observational study in which information is collected on subjects' past experience. A case-control study more specifically involves a sample of individuals having a particular disease as well as a sample of individuals not having the disease, and collects information about the individuals' past exposure to putative causes of the disease.

Risk ratio See relative risk.

RMS (root mean squared) error To calculate the RMS (root mean squared) error the individual errors are squared, added together, divided by the number of individual errors, and then square rooted. Gives a single number which summarizes the overall error.

RMS error of regression The rms error of regression is the rms of the vertical residuals from the regression line. For regressing Y on X, the rms error of regression is equal to $(1 - r^2)^{1/2} \times SD_Y$, where r is the correlation coefficient between X and Y and SD_Y is the standard deviation of the values of Y.

RNG Acronym for random number generator. This is a process which uses a arithmetic algorithm to generate sequences of pseudo-random numbers. Also See seed.

Robust Describing a statistic (or parameter) that is little affected by outliers or gross errors of observation.

Root cause analysis The term root cause analysis is commonly used in manufacturing to summarize the activities involved in determining the variables or factors that impact the final quality or yield of the respective processes. For example, if a particular pattern of defects emerges in manufacture of silicon chips, engineers will pursue various methods and strategies for root cause analysis to determine the ultimate causes of those patterns of quality problems.

Root mean square (RMS) The RMS of a list is the square-root of the mean of the squares of the elements in the list. It is a measure of the average "size" of the elements of the list. To compute the RMS of a list, you square all the entries, average the numbers you get, and take the square-root of that average.

Root mean square error (RMSE) The RMSE of an an estimator of a parameter is the square-root of the mean squared error (MSE) of the estimator. In symbols, if X is an estimator of the parameter t, then

$$RMSE(X) = (E((X-t)^2))^{1/2}.$$

The RMSE of an estimator is a measure of the expected error of the estimator. The units of RMSE are the same as the units of the estimator. See also mean squared error.

Root mean square standardized effect (RMSSE) This standardized measure of effect size is used in the Analysis of Variance to characterize the overall level of population effects. It is the square root of the sum of squared standardized effects divided by the number of degrees of freedom for the effect. For example, in a 1-Way Anova, the RMSSE is calculated as

$$RMSSE = \sqrt{\frac{\sum_{i=1}^{n}\left(\frac{\alpha_j}{\sigma}\right)}{J-1}} = \sqrt{\frac{S_\alpha^2}{\sigma^2}} = \frac{S_\alpha}{\sigma}$$

Rosenbrock pattern search This Nonlinear Estimation method will rotate the parameter space and align one axis with a ridge (this method is also called the method of rotating coordinates); all other axes will remain orthogonal to this axis. If the loss function is unimodal and has detectable ridges pointing towards the minimum of the function, then this method will proceed with a considerable accuracy towards the minimum of the function.

Rounding A method of simplifying numbers to fewer significant figures than in the original number. There are several ways to round numbers but the most common is to round down a number ending in 1 through 4 and round up numbers ending in 5 through 9. For example, 41.4 rounds down to 41 and 41.5 rounds up to 42.

Row percentages A percentage calculated by finding the count in a cell compared with the total count for the row (the row marginal count). For example, if there are 530 unemployed women in a sample including 648 unemployed people in all, the row percentage of unemployed women is 530/648 %, or 82%.

Running medians smoothing Running medians smoothing is a smoothing technique analogous to that used for moving averages. The purpose of the technique is the same, to make a trend clearer by reducing the effects of other fluctuations.

Runs test In studies where measurements are made according to some well defined ordering, either in time or space, a frequent question is whether or not the average value of the measurement is different at different points in the sequence. The runs test provides a means of testing this.

Example

Suppose that, as part of a screening programme for heart disease, men aged 45-65 years have their blood cholesterol level measured on entry to the study. After many months it is noticed that cholesterol levels in this population appear somewhat higher in the Winter than in the Summer. This could be tested formally using a Runs test on the recorded data, first arranging the measurements in the date order in which they were collected.

Runs tests (in quality control) These tests are designed to detect patterns measurement (e.g., sample means) that may indicate that the process is out of control. In quality control charting, when a sample point (e.g., a mean in an X-bar chart) falls outside the control lines, one has reason to believe that the process may no longer be in control. In addition, one should look for systematic patterns of points (e.g., means) across samples, because such patterns may indicate that the process average has shifted. Most quality control software packages will (optionally) perform the standard set of tests for such patterns; these tests are also sometimes referred to as AT&T runs rules or tests for special causes. The term special or assignable causes as opposed to chance or common causes was used by Shewhart to distinguish between a process that is in control, with variation due to random (chance) causes only, from a process that is out of control, with variation that is due to some non-chance or special (assignable) factors.

As the sigma control limits for quality control charts, the runs rules are based on "statistical" reasoning. For example, the probability of any sample mean in an X-bar control chart falling above the center line is equal to 0.5, provided 1. That the process is in control (i.e., that the center line value is equal to the population mean),

2. That consecutive sample means are independent (i.e., not auto-correlated), and

3. That the distribution of means follows the normal distribution. Simply stated, under those conditions there is a 50-50 chance that a mean will fall above or below the center line. Thus, the probability that two consecutive means will fall above the center line is equal to 0.5 times 0.5 = 0.25.

S

S.D. ratio In a regression problem, the ratio of the prediction error standard deviation to the original output data standard deviation. A lower S.D. ratio indicates a better prediction. This is equivalent to one minus the explained variance of the model. See Multiple Regression.

Sample A sample is a group of units selected from a larger group (the population). By studying the sample it is hoped to draw valid conclusions about the larger group.

A sample is generally selected for study because the population is too large to study in its entirety. The sample should be representative of the general population. This is often best achieved by random sampling. Also, before collecting the sample, it is important that the researcher carefully and completely defines the population, including a description of the members to be included.

Example

The population for a study of infant health might be all children born in the U.K. in the 1980's. The sample might be all babies born on 7th May in any of the years. A set of individual units, drawn from some definable population of units, and generally a small proportion of the population, to be used for a statistical examination of which the findings are intended to be applied to the population. It is essential for such inference that the sample should be representative. In re-randomisation statistics the process of applying inferences based upon the sample to the population is essentially informal.

Sample distribution The frequency distribution of an empirical sample. Not to be confused with sampl*ing* distributions.

Sample mean The arithmetic mean of a random sample from a population. It is a statistic commonly used to estimate the population mean. Suppose there are n data, $\{x_1, x_2, \ldots, x_n\}$. The sample mean is $(x_1 + x_2 + \ldots + x_n)/n$. The expected value of the sample mean is the population mean. For sampling with replacement, the SE of the sample mean is the population standard deviation, divided by the square-root of the sample size. For sampling without replacement, the SE of the sample mean is the finite-population correction $((N\text{-}n)/(N\text{-}1))^{½}$ times the SE of the sample mean for sampling with replacement, with N the size of the population and n the size of the sample. The sample mean is an estimator available for estimating the population mean . It is a measure of location, commonly called the average, often symbolised $\overline{x}$

Example

Lets say our data set is: 5 3 54 93 83 22 17 19.

The sample mean is calculated by taking the sum of all the data values and dividing by the total number of data values:

$$X\bar{x} = \frac{5+3+54+93+83+22+17+19}{8} = 37$$

Its value depends equally on all of the data which may include outliers. It may not appear representative of the central region for skewed data sets.

It is especially useful as being representative of the whole sample for use in subsequent calculations.

Sample percentage The percentage of a random sample with a certain property, such as the percentage of voters registered as Democrats in a simple random sample of voters. The sample mean is a statistic commonly used to estimate the population percentage. The expected value of the sample percentage from a simple random sample or a random sample with replacement is the population percentage. The SE of the sample percentage for sampling with replacement is $(p(1-p)/n)^{1/2}$, where p is the population percentage and n is the sample size. The SE of the sample percentage for sampling without replacement is the finite-population correction $((N-n)/(N-1))^{1/2}$ times the SE of the sample percentage for sampling with replacement, with N the size of the population and n the size of the sample. The SE of the sample percentage is often estimated by the bootstrap.

Sample size The number of elements in a sample from a population. The number of experimental units on which observations are considered. This may be less than the number of observations in a data-set, due to the possible multipying effects of multiple variables and/or repeated measures within the experimental design.

Sample space The sample space is an exhaustive list of all the possible outcomes of an experiment. Each possible result of such a study is represented by one and only one point in the sample space, which is usually denoted by S.

Examples

1. Experiment Rolling a die once:

Sample space S = {1,2,3,4,5,6}

2. Experiment Tossing a coin:

Sample space S = {Heads,Tails}

3. Experiment Measuring the height (cms) of a girl on her first day at school:

Sample space S = the set of all possible real numbers

Sample standard deviation, *s* The sample standard deviation *S* is an estimator of the standard deviation of a population based on a random sample from the population. The sample standard deviation is a statistic that measures how "spread out" the sample is around the sample mean. It is quite similar to the standard deviation of the sample, but instead of averaging the squared deviations (to get the rms of the deviations of the data from the sample mean) it divides the sum of the squared deviations by (number of data – 1) before taking the square-root. Suppose there are n data, $\{x_1, x_2, \ldots, x_n\}$, with mean $M = (x_1 + x_2 + \ldots + x_n)/n$. Then

$s = (((x_1 - M)^2 + (x_2 - M)^2 + \ldots + (x_n - M)^2)/(n-1))^{1/2}$

The square of the sample standard deviation, S^2 (the sample variance) is an unbiased estimator of the square of the SD of the population (the variance of the population).

Sample sum The sum of a random sample from a population. The expected value of the sample sum is the sample size times the population mean. For sampling with replacement, the SE of the sample sum is the population standard deviation, times the square-root of the sample size. For sampling without replacement, the SE of the sample sum is the finite-population correction $((N-n)/(N-1))^{1/2}$ times the SE of the sample sum for sampling with replacement, with N the size of the population and n the size of the sample.

Sample survey A survey based on the responses of a sample of individuals, rather than the entire population.

Sample variance Sample variance is a measure of the spread of or dispersion within a set of sample data.

The sample variance is the sum of the squared deviations from their average divided by one less than the number of observations in the data set. For example, for n observations x1, x2, x3,....., xn with sample mean:

$$\bar{x} = \frac{\sum x_i}{n}$$

the sample variance is:

$$s^2 = \frac{1}{n-1}\sum (x_i - \bar{x})^2$$

See also variance.

Sampling distribution A theoretical frequency distribution of any statistic calculated for a sample. For example, the sampling distribution of the mean could be generated by collecting an infinite number of similar size samples and calculating the means for each. These means would then form a distribution which could be plotted and would form a normal curve. Sampling distributions are the basis of inferential statistics. The sampling distribution describes probabilities associated with a statistic when a random sample is drawn from a population.

The sampling distribution is the probability distribution or probability density function of the statistic.

Derivation of the sampling distribution is the first step in calculating a confidence interval or carrying out a hypothesis test for a parameter.

Sampling distribution The sampling distribution of an estimator is the probability distribution of the estimator when it is applied to random samples.

Sampling error In estimating from a random sample, the difference between the estimator and the parameter can be written as the sum of two components: bias and sampling error. The bias is the average error of the estimator over all possible samples. The bias is not random. Sampling error is the component of error that varies from sample to sample. The sampling error is random: it comes from "the luck of the draw" in which units happen to be in the sample. It is the chance variation of the estimator. The average of the sampling error over all possible samples (the expected value of the sampling error) is zero. The standard error of the estimator is a measure of the typical size of the sampling error.

Sampling unit A sample from a population can be drawn one unit at a time, or more than one unit at a time (one can sample clusters of units). The fundamental unit of the sample is called the *sampling unit*. It need not be a unit of the population.

Sampling variability Sampling variability refers to the different values which a given function of the data takes when it is computed for two or more samples drawn from the same population.

Saturated A model that includes all possible effects.

Saturated model A model for data with a binary response variable in which a separate probability is estimated for every observed set of conditions (as described by explanatory variables).

Scalable software systems Software (e.g., a data base management system, such as MS SQL Server or Oracle) that can be expanded to meet future requirements without the need to restructure its operation (e.g., split data into smaller segments) to avoid a degradation of its performance. For example, a scalable network allows the network administrator to add many additional nodes without the need to redesign the basic system. An example of a non-scalable architecture is the DOS directory structure (adding files will eventually require splitting them into subdirectories).

Scale type See measurement type.

Scaling Altering original variable values (according to a specific function or an algorithm) into a range that meet particular criteria (e.g., positive numbers, fractions, numbers less than 10E12, numbers with a large relative variance).

Scatterplot A scatterplot is a way to visualize bivariate data. A scatterplot is a plot of pairs of measurements on a collection of "individuals" (which need not be people). For example, suppose we record the heights and weights of a group of 100 people. The scatterplot of those data would be 100 points. Each point represents one person's height and weight. In a scatterplot of weight *against* height, the *x*-coordinate of each point would be height of one person, the *y*-coordinate of that point would be the weight of the same person. In a scatterplot of height against weight, the *x*-coordinates would be the weights and the *y*-coordinates would be the heights.

Scatterplot smoothers In 2D scatterplots, various smoothing methods are available to fit a function through the points to best represent (summarize) the relationship between the variables.

Scatterplot, 2D The scatterplot visualizes a relation (correlation) between two variables X and Y (e.g., weight and height). Individual data points are represented in two-dimensional space, where axes represent the variables (X on the horizontal axis and Y on the vertical axis).

The two coordinates (X and Y) that determine the location of each point correspond to its specific values on the two variables.

See also, Data Reduction.

Scatterplot, 2D categorized ternary graph The points representing the proportions of the component variables (X, Y, and Z) in a ternary graph are plotted in a 2-dimensional display for each level of the grouping variable (or user-defined subset of data). One component graph is produced for each level of the grouping variable (or user-defined subset of data) and all the component graphs are arranged in one display to allow for comparisons between the subsets of data (categories).

See also, Data Reduction.

Scatterplot, 2D double Y This type of scatterplot can be considered to be a combination of two multiple scatterplots for one X-variable and two different sets (lists) of Y-variables. A scatterplot for the X-variable and each of the selected Y-variables will be plotted, but the variables entered into the first list (called Left-Y) will be plotted against the left-Y axis, whereas the variables entered into the second list (called Right-Y) will be plotted against the right-Y axis. The names of all Y-variables from the two lists will be included in the legend followed either by the letter (L) or (R), denoting the left-Y and right-Y axis, respectively.

The Double-Y scatterplot can be used to compare images of several correlations by overlaying them in a single graph. However, due to the independent scaling used for the two list of variables, it can facilitate comparisons between variables with values in different ranges.

Scatterplot, 2D frequency Frequency scatterplots display the frequencies of overlapping points between two variables in order to visually represent data point weight or other measurable characteristics of individual data points.

See also, Data Reduction.

Scatterplot, 2D multiple Unlike the regular scatterplot in which one variable is represented by the horizontal axis and one by the vertical axis, the multiple scatterplot consists of multiple plots and represents multiple correlations: one variable (X) is represented by the horizontal axis, and several variables (Y's) are plotted against the vertical axis. A different point marker and color is used for each of the multiple Y-variables and referenced in the legend so that individual plots representing different variables can be discriminated in the graph.

The Multiple scatterplot is used to compare images of several correlations by overlaying them in a single graph that uses one common set of scales (e.g., to reveal the underlying structure of factors or dimensions in Discriminant Function Analysis).

See also, Data Reduction.

Scatterplot, 2D regular The regular scatterplot visualizes a relation between two variables X and Y (e.g., weight and height). Individual data points are represented by point markers in two- dimensional space, where axes represent the variables. The two coordinates (X and Y) which determine the location of each point, correspond to its specific values on the two variables. If the two variables are strongly related, then the data points form a systematic shape (e.g., a straight line or a clear curve). If the variables are not related, then the points form an irregular "cloud."

Fitting functions to scatterplot data helps identify the patterns of relations between variables.

Scatterplot, 3D 3D Scatterplots visualize a relationship between three or more variables, representing the X, Y, and one or more Z (vertical) coordinates of each point in 3-dimensional space.

See also, 3D Scatterplot - Custom Ternary Graph, Data Reduction and Data Rotation (in 3D space).

Scatterplot, 3D raw data An unsmoothed surface (no smoothing function is applied) is drawn through the points in the 3D scatterplot.

See also, Data Reduction.

Scatterplot, 3D ternary graph In this type of ternary graph, the triangular coordinate systems are used to plot four (or more) variables (the components X, Y, and Z, and the responses V1, V2, etc.) in three dimensions (ternary 3D scatterplots or surface plots). Here, the responses (V1, V2, etc.) associated with the proportions of the component variables (X, Y, and Z) in a ternary graph are plotted as the heights of the points.

See also, Data Reduction.

Scheffe's test This post hoc test can be used to determine the significant differences between group means in an analysis of variance setting. Scheffe's test is considered to be one of the most conservative post hoc tests. See also, Post Hoc Comparisons.

Score statistic This statistic is used to evaluate the statistical significance of parameter estimates computed via maximum likelihood methods. It is also sometimes called the efficient score statistic. The test is based on the behavior of the log-likelihood function at the point where the respective parameter estimate is equal to 0.0 (zero); specifically, it uses the derivative (slope) of the log-likelihood function evaluated at the null hypothesis value of the parameter (parameter = 0.0). While this test is not as accurate as explicit likelihood-ratio test statistics based on the ratio of the likelihoods of the model that includes the parameter of interest, over the likelihood of the model that does not, its computation is usually much faster. It is therefore the preferred method for evaluating the statistical significance of parameter estimates in stepwise or best-subset model building methods.

An alternative statistic is the Wald statistic.

Scree plot, scree test The eigenvalues for successive factors can be displayed in a simple line plot. Cattell (1966) proposed that this scree plot can be used to graphically determine the optimal number of factors to retain.

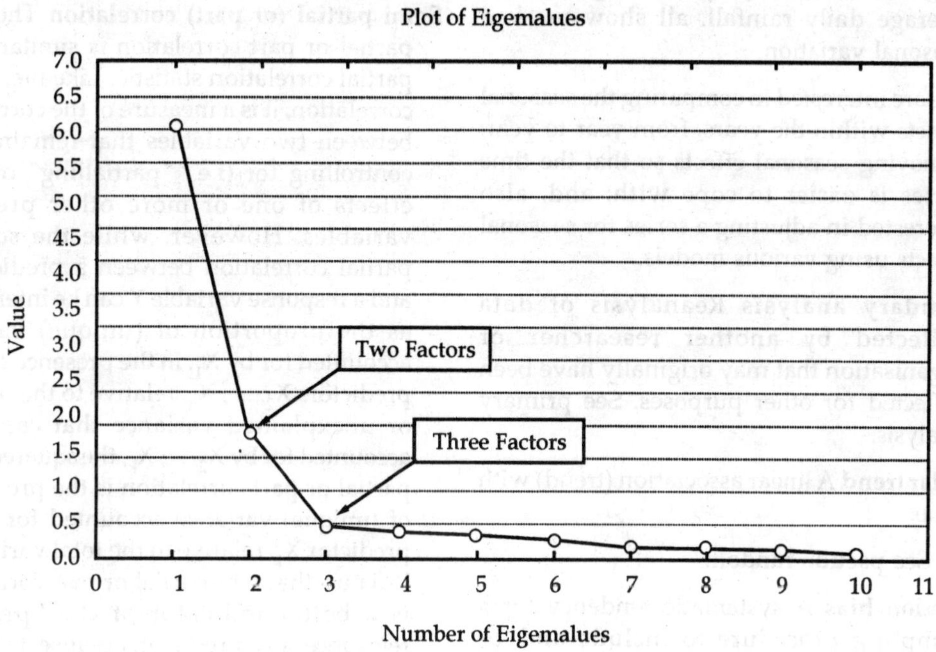

The scree test involves finding the place where the smooth decrease of eigenvalues appears to level off to the right of the plot. To the right of this point, presumably, one finds only "factorial scree" — "scree" is the geological term referring to the debris which collects on the lower part of a rocky slope. Thus, no more than the number of factors to the left of this point should be retained.

SD line For a scatterplot, a line that goes through the point of averages, with slope equal to the ratio of the standard deviations of the two plotted variables. If the variable plotted on the horizontal axis is called X and the variable plotted on the vertical axis is called Y, the slope of the SD line is the SD of Y, divided by the SD of X.

Seasonal component We want to increase our understanding of a time series by picking out its main features. One of these main features is the seasonal component. Descriptive techniques may be extended to forecast (predict) future values.

In weekly or monthly data, the seasonal component, often referred to as seasonality, is the component of variation in a time series which is dependent on the time of year. It describes any regular fluctuations with a period of less than one year. For example, the costs of various types of fruits and vegetables, unemployment figures and average daily rainfall, all show marked seasonal variation.

We are interested in comparing the seasonal effects within the years, from year to year; removing seasonal effects so that the time series is easier to cope with; and, also interested in adjusting a series for seasonal effects using various models.

Secondary analysis Reanalysis of data collected by another researcher or organisation that may originally have been collected for other purposes. See primary analysis.

Secular trend A linear association (trend) with time.

Seed See pseudo-random.

Selection bias A systematic tendency for a sampling procedure to include and/or exclude units of a certain type. For example, in a quota sample, unconscious prejudices or predilections on the part of the interviewer can result in selection bias. Selection bias is a potential problem whenever a human has latitude in selecting individual units for the sample; it tends to be eliminated by probability sampling schemes in which the interviewer is told exactly whom to contact (with no room for individual choice).

Self selection Self-selection occurs when individuals decide for themselves whether they are in the control group or the treatment group. Self-selection is quite common in studies of human behavior. For example, studies of the effect of smoking on human health involve self-selection: individuals choose for themselves whether or not to smoke. Self-selection precludes an experiment; it results in an observational study. When there is self-selection, one must be wary of possible confounding from factors that influence individuals' decisions to belong to the treatment group.

Semi partial (or part) correlation The semi-partial or part correlation is similar to the partial correlation statistic. Like the, partial correlation, it is a measure of the correlation between two variables that remains after controlling for (i.e., "partialling" out) the effects of one or more other predictor variables. However, while the squared partial correlation between a predictor X_1 and a response variable Y can be interpreted as the proportion of (unique) variance accounted for by X_1, in the presence of other predictors $X_2, \ldots, X_k$, relative to the residual or unexplained variance that cannot be accounted for by $X_2, \ldots, X_k$, the squared semi-partial or part correlation is the proportion of (unique) variance accounted for by the predictor X_1, relative to the total variance of Y. Thus, the semi-partial or part correlation is a better indicator of the "practical relevance" of a predictor, because it is scaled to (i.e., relative to) the total variability in the dependent (response) variable.

See also Correlation, Spurious Correlations, partial correlation, Basic Statistics, Multiple Regression, General Linear Models, General Stepwise Regression, Structural Equation Modeling (SEPATH).

Sensitivity analysis A sensitivity analysis indicates which input variables are considered most important by that particular neural network. Sensitivity analysis can be used purely for informative purposes, or to perform input pruning.

Sensitivity analysis can give important insights into the usefulness of individual variables. It often identifies variables that can be safely ignored in subsequent analyses, and key variables that must always be retained. However, it must be deployed with some care, for reasons that are explained below.

Input variables are not, in general, independent - that is, there are interdependencies between variables. Sensitivity analysis rates variables according to the deterioration in modeling performance that occurs if that variable is no longer available to the model. In so doing, it assigns a single rating value to each variable. However, the interdependence between variables means that no scheme of single ratings per variable can ever reflect the subtlety of the true situation.

Consider, for example, the case where two input variables encode the same information (they might even be copies of the same variable). A particular model might depend wholly on one, wholly on the other, or on some arbitrary combination of them. Then sensitivity analysis produces an arbitrary relative sensitivity to them. Moreover, if either is eliminated the model may compensate adequately because the other still provides the key information. It may therefore rate the variables as of low sensitivity, even though they might encode key information. Similarly, a variable that encodes relatively unimportant information, but is the only variable to do so, may have higher sensitivity than any number of variables that mutually encode more important information.

There may be interdependent variables that are useful only if included as a set. If the entire set is included in a model, they can be accorded significant sensitivity, but this does not reveal the interdependency. Worse, if only part of the interdependent set is included, their sensitivity will be zero, as they carry no discernable information.

In summary, sensitivity analysis does not rate the "usefulness" of variables in modeling in a reliable or absolute manner. You must be cautious in the conclusions you draw about the importance of variables. Nonetheless, in practice it is extremely useful. If a number of models are studied, it is often possible to identify key variables that are always of high sensitivity, others that are always of low sensitivity, and "ambiguous" variables that change ratings and probably carry mutually redundant information.

Separate variance t test A t test for two samples that uses the variance estimated from the cases in each sample separately. A separate variance t test is appropriate when the samples are drawn from the different populations.

Sequential contour plot, 3D This contour plot presents a 2-dimensional projection of the spline-smoothed surface fit to the data. Successive values of each series are plotted along the X-axis, with each successive series represented along the Y- axis.

Sequential surface plot, 3D In this sequential plot, a spline-smoothed surface is fit to each data point. Successive values of each series are plotted along the X-axis, with each successive series represented along the Y-axis.

Sequential/stacked plots In this type of graph, the sequence of values from each selected variable is stacked on one another.

Sequential/stacked plots, 2D area The sequence of values from each selected variable will be represented by consecutive areas stacked on one another in this type of graph.

Sequential/stacked plots, 2D column The sequence of values from each selected variable will be represented by consecutive segments of vertical columns stacked on one another in this type of graph.

Sequential/stacked plots, 2D lines The sequence of values from each selected variable will be represented by consecutive lines stacked on one another in this type of graph.

Sequential/stacked plots, 2D mixed line In this type of graph, the sequences of values of variables selected in the first list will be represented by consecutive areas stacked on one another while the sequences of values of variables selected in the second list will be represented by consecutive lines stacked on one another (over the area representing the last variable from the first list).

Sequential/stacked plots, 2D mixed step In this type of graph, the sequences of values of variables selected in the first list will be represented by consecutive step areas stacked on one another while the sequences of values of variables selected in the second list will be represented by consecutive step lines stacked on one another (over the step area representing the last variable from the first list).

Sequential/stacked plots, 2D step area The sequence of values from each selected variable will be represented by consecutive step areas stacked on one another in this type of graph.

Sequential/stacked plots, 2D step The sequence of values from each selected variable will be represented by consecutive step lines stacked on one another in this type of graph.

Set A set is a collection of things, without regard to their order.

Sets of samples in quality control charts While monitoring an ongoing process, it often becomes necessary to adjust the center line values or control limits, as those values are being refined over time. Also, one may want to compute the control limits and center line values from a set of samples that are known to be in control, and apply those values to all subsequent samples. Thus, each set is defined by a set of computation samples (from which various statistics are computed, e.g., sigma, means, etc.) and a set of application samples (to which the respective statistics, etc. are applied). Of course, the computation samples and application samples can be (and often are) not the same. To reiterate, you may want to estimate sigma from a set of samples that are known to be in control (the computation set), and use that estimate for establishing control limits for all remaining and new samples (the application set).

Note that each sample must be uniquely assigned to one application set; in other words, each sample has control limits based on statistics (e.g., sigma) computed for one particular set. The assignment of application samples to sets proceeds in a hierarchical manner, i.e., each sample is assigned to the first set where it "fits" (where the definition of the application sample set would include the respective sample). This hierarchical search always begins at the last set that the user specified, and not with the all-samples set. Hence, if the user-specified sets encompass all valid samples, the default all-samples set will actually become empty (since all samples will be assigned to one of the user-defined sets).

Shapiro Wilk's W test The Shapiro-Wilks' W test is used in testing for normality. If the W statistic is significant, then the hypothesis that the respective distribution is normal should be rejected. The Shapiro-Wilks' W test is the preferred test of normality because of its good power properties as compared to a wide range of alternative tests (Shapiro, Wilk, & Chen, 1968). Some software programs implement an extension to the test described by Royston (1982), which allows it to be applied to large samples (with up to 5000 observations). See also Kolmogorov-Smirnov test and Lilliefors test.

Shewhart control charts This is a standard graphical tool widely used in statistical Quality Control. The general approach to quality control charting is straightforward: One extracts samples of a certain size from the ongoing production process. One then produces line charts of the variability in those samples, and consider their closeness to target specifications. If a trend emerges in those lines, or if samples fall outside pre-specified limits, then the process is declared to be out of control and the operator will take action to find the cause of the problem. These types of charts are sometimes also referred to as Shewhart control charts (named after W. A. Shewhart who is generally credited as being the first to introduce these methods.

For additional information, see also Quality Control charts; Assignable causes and actions.

Shift An ongoing process that at some point was centered will shift over time. Motorola, in their implementation of Six Sigma strategies, determined that it is reasonable to assume that a process will shift over time by approximately 1.5 * sigma. Hence, most standard Six Sigma calculators will be based on a 1.5 * sigma shift.

Shift algorithm Algorithms employing branch-and-bound methods for the Pitman permutaion test(1) and the Pitman permutation test(2).

Short run control charts The short run quality control chart , for short production runs, plots transformations of the observations of variables or attributes for multiple parts, each of which constitutes a distinct "run," on the same chart. The transformations rescale the variable values of interest such that they are of comparable magnitudes across the different short production runs (or parts). The control limits computed for those transformed values can then be applied to determine if the production process is in control, to monitor continuing production, and to establish procedures for continuous quality improvement.

Shuffle data Randomly assigning cases to the training and verification sets, so that these are (as far as possible) statistically unbiased.

Shuffle, back propagation Presenting training cases in a random order on each epoch, to prevent various undesirable effects which can otherwise occur (such as oscillation and convergence to local minima).

Sigma restricted model A sigma restricted model uses the sigma-restricted coding to represent effects for categorical predictor variables in general linear models and generalized linear models. To illustrate the sigma-restricted coding, suppose that a categorical predictor variable called Gender has two levels (i.e., male and female). Cases in the two groups would be assigned values of 1 or -1, respectively, on the coded predictor variable, so that if the regression coefficient for the variable is positive, the group coded as 1 on the predictor variable will have a higher predicted value (i.e., a higher group mean) on the dependent variable, and if the regression coefficient is negative, the group coded as -1 on the predictor variable will have a higher predicted value on the dependent variable. This coding strategy is aptly called the sigma-restricted parameterization, because the values used to represent group membership (1 and -1) sum to zero.

Sigmoid function An S-shaped curve, with a near-linear central response and saturating limits.

Sign test The sign test is designed to test a hypothesis about the location of a population distribution. It is most often used to test the hypothesis about a population median, and often involves the use of matched pairs, for example, before and after data, in which case it tests for a median difference of zero.

The Sign test does not require the assumption that the population is normally distributed.

In many applications, this test is used in place of the one sample t-test when the normality assumption is questionable. It is a less powerful alternative to the Wilcoxon signed ranks test, but does not assume that the population probability distribution is symmetric.

This test can also be applied when the observations in a sample of data are ranks, that is, ordinal data rather than direct measurements.

Signal detection theory (SDT) Signal detection theory (SDT) is an application of statistical decision theory used to detect a signal embedded in noise. SDT is used in psychophysical studies of detection, recognition, and discrimination, and in other areas such as medical research, weather forecasting, survey research, and marketing research.

A general approach to estimating the parameters of the signal detection model is via the use of the generalized linear model. For example, DeCarlo (1998) shows how signal detection models based on different underlying distributions can easily be considered by using the generalized linear model with different link functions.

Significance See statistical significance.

Significance level The significance level of a statistical hypothesis test is a fixed probability of wrongly rejecting the null hypothesis H0, if it is in fact true.

It is the probability of a type I error and is set by the investigator in relation to the consequences of such an error. That is, we want to make the significance level as small as possible in order to protect the null hypothesis and to prevent, as far as possible, the investigator from inadvertently making false claims.

The significance level is usually denoted by α

Significance Level = P(type I error) = α

Usually, the significance level is chosen to be = 0.05 = 5%.

Significance test A procedure for assessing the evidence in a set of data against a null hypothesis. The result of a signficance test is a P-value. cf. hypothesis test.

Simple hypothesis A simple hypothesis is a hypothesis which specifies the population distribution completely.

Examples

1. H0: X~Bi(100,1/2) i.e. p is specified.

2. H0: X~N(5,20) i.e. μ and σ^2 are specified.

See also composite hypothesis.

Simple linear regression Simple linear regression aims to find a linear relationship between a response variable and a possible predictor variable by the method of least squares.

Simple random sample A simple random sample of *n* units from a population is a random sample drawn by a procedure that is equally likely to give every collection of *n* units from the population; that is, the probability that the sample will consist of any given subset of *n* of the *N* units in the population is $1/{}_{N}C_{n}$. Simple random sampling is sampling at random without replacement (without replacing the units between draws). A simple random sample of size n from a population of $N >= n$ units can be constructed by assigning a random number between zero and one to each unit in the population, then taking those units that were assigned the n largest random numbers to be the sample.

Simple random sampling (SRS) Simple random sampling is a type of probability sampling where observations are randomly selected from a population with a known probability or sampling fraction. Typically, one begins with a list of N observations that comprises the entire population from which one wishes to extract a simple random sample (e.g., a list of registered voters); one can then generate k random case numbers (without replacement) in the range from 1 to N, and select the respective cases into the final sample (with a sampling fraction or known selection probability of k/N). Simple random sampling is the basic sampling technique where we select a group of subjects (a sample) for study from a larger group (a population). Each individual is chosen entirely by chance and each member of the population has an equal chance of being included in the sample. Every possible sample of a given size has the same chance of selection; that is, each member of the population is equally likely to be chosen at any stage in the sampling process.

Compare random sampling.

Simple regression analysis Statistical methods concerned with explaining or predicting the variability of a continuous dependent variable using information from one continuous independent variable.

Simplex algorithm A nonlinear estimation algorithm that does not rely on the computation or estimation of the derivatives of the loss function. Instead, at each iteration the function will be evaluated at m+1 points in the m dimensional parameter space. For example, in two dimensions (i.e., when there are two parameters to be estimated), the program will evaluate the function at three points around the current optimum. These three points would define a triangle; in more than two dimensions, the "figure" produced by these points is called a Simplex.

Simpson's paradox The phenomenon that either assocation or lack of association between variables within strata may be eliminated or reversed when the observations are pooled across strata. Briefly, the fact that neither one of conditional independence nor marginal independence implies the other.

Sine Possibly useful if recognizing radially-distributed data; not used by default.

Single and multiple censoring There are situations in which censoring can occur at different times (multiple censoring), or only at a particular point in time (single censoring). Consider an example experiment where we start with 100 light bulbs, and terminate the experiment after a certain amount of time. If the experiment is terminated at a particular point in time, then a single point of censoring exists, and the data set is said to be single-censored. However, in biomedical research multiple censoring often exists, for example, when patients are discharged from a hospital after different amounts (times) of treatment, and the researcher knows that the patient survived up to those (differential) points of censoring.

Data sets with censored observations can be analyzed via Survival Analysis or via Weibull and Reliability/Failure Time Analysis. See also, Type I and II Censoring and Left and Right Censoring.

Single output networks Levenberg-Marquardt can only be used on networks with a single output unit.

Single output neuron Two thresholds are used: accept and reject. In the single output neuron case, the output is considered to be the first class if the output neuron's activation is below the reject threshold, and to be the second class if its activation is above the accept threshold. If the activation is between the two thresholds, the class is regarded as "unknown" (the so-called doubt option). If the two thresholds are equal, there is no doubt option. There are two common configurations: accept=reject=0.5 implies no doubt option, with the most likely class assigned; accept=0.95, reject=0.05 implies standard "95% confidence" in assignment of a class, with doubt expressed otherwise. Both of these cases assume the standard logistic output neuron activation function, which gives the output neuron a (0,1) range; the thresholds should be adjusted accordingly for different activation functions [e.g., hyperbolic tangent uses output range (-1,+1)].

As an alternative to selecting the thresholds yourself, you can specify a loss coefficient that gives the relative "cost" of the two possible misclassifications (false-positive versus false-negative). A loss coefficient of 1.0 indicates that the two classes are equally important. A loss coefficient above 1.0 indicates that it is relatively more important to correctly recognize class two cases, even at the expense of misclassifying more class one cases. The thresholds (which are equal) are determined by calculating a ROC curve and determining the point on the curve where the ratio of false positives to false negatives equals the loss coefficient. This equalizes the weighted loss on each class, independent of the number of cases in each (i.e. with a loss coefficient of 1.0, it is the proportion of misclassifications in each class that is equalized, not the number of absolute number of misclassifications).

Singular value decomposition An efficient algorithm for optimizing a linear model.

Six sigma (DMAIC) Six Sigma is a well-structured, data-driven methodology for eliminating defects, waste, or quality control problems of all kinds in manufacturing, service delivery, management, and other business activities. Six Sigma methodology is based on the combination of well-established statistical quality control techniques, simple and advanced data analysis methods, and the systematic training of all personnel at every level in the organization involved in the activity or process targeted by Six Sigma.

Six Sigma methodology and management strategies provide an overall framework for organizing company wide quality control efforts. These methods have recently become very popular, due to numerous success stories from major US-based as well as international corporations.

These are organized into the categories of activities that make up the Six Sigma effort: Define (D), Measure (M), Analyze (A), Improve (I), Control (C); or dmaic for short.

Six sigma process A six sigma process is one that can be expected to produce only 3.4 defects per one million opportunities. The concept of the six sigma process is important in Six Sigma quality improvement programs. The idea can best be summarized with the following graphs.

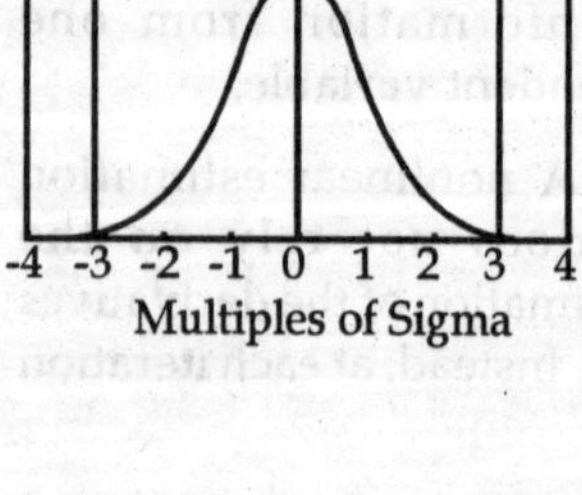

Six Sigma Process —Standard 3 Sigma process

We start with a standard 3-Sigma process. The upper and lower specification limits are set at ± 3 * Sigma

The term Six Sigma derives from the goal to achieve a process variation, so that ± 6 * sigma (the estimate of the population standard deviation) will "fit" inside the lower and upper specification limits for the process. In that case, even if the process mean shifts by 1.5 * sigma in one direction (e.g., to +1.5 sigma in the direction of the upper specification limit), then the process will still produce very few defects.

For example, suppose we expressed the area above the upper specification limit in terms of one million opportunities to produce defects. The 6 * sigma process shifted upwards by 1.5 * sigma will only produce 3.4 defects (i.e., "parts" or "cases" greater than the upper specification limit) per one million opportunities.

Size (of a sample) The number of individuals in the sample.

(of a hypothesis test) the probability of a type I error; or more exactly, the least upper bound on the probability of a type I error.

Skewed A distribution displaced at one end of the scale and a tail strung out at the other end.

Skewed distribution A distribution which has either predominately low values and a few extreme high values (positively skewed) or one where there are predominately high values and a few extreme low values (negatively skewed). When plotted, such a distribution produces a non-symmetric curve where the mean, mode and median do not coincide.

Skewness Skewness measures the deviation of the distribution from symmetry. If the skewness is clearly different from 0, then that distribution is asymmetrical, while normal distributions are perfectly symmetrical.

$$Skewness = n*M_3/[(n-1)*(n-2)*\sigma^3]$$

Where M_3 is equal to: $\Sigma\ (x_i\text{-Mean}_x)^3$

σ^3 is the standard deviation (sigma) raised to the third power

n is the valid number of cases. Skewness is defined as asymmetry in the distribution of the sample data values. Values on one side of the distribution tend to be further from the 'middle' than values on the other side.

For skewed data, the usual measures of location will give different values, for example, mode<median<mean would indicate positive (or right) skewness. Positive (or right) skewness is more common than negative (or left) skewness.

If there is evidence of skewness in the data, we can apply transformations, for example, taking logarithms of positive skew data.

Compare symmetry.

Slope A measure of a line's slant.

Small networks Levenberg-Marquardt has space requirements proportional to the square of the number of weights in the network. This effectively precludes its use in networks of any great size (more than a few hundred weights).

Smoothing Smoothing techniques can be used in two different situations. Smoothing techniques for 3D Bivariate Histograms allow you to fit surfaces to 3D representations of bivariate frequency data. Thus, every 3D histogram can be turned into a smoothed surface providing a sensitive method for revealing non-salient overall patterns of data and/or identifying patterns to use in developing quantitative models of the investigated phenomenon.

In Time Series analysis, the general purpose of smoothing techniques is to "bring out" the major patterns or trends in a time series, while de-emphasizing minor fluctuations (random noise). Visually, as a result of smoothing, a jagged line pattern should be transformed into a smooth curve.

See also, Exploratory Data Analysis and Data Mining Techniques, and Smoothing Bivariate Distributions.

Softmax A specialized activation function for one-of-N encoded classification networks. Performs a normalized exponential (i.e. the outputs add up to 1). In combination with the cross entropy error function, allows multilayer perceptron networks to be modified for class probability estimation

Softmax Exponential function, with results normalized so that the sum of activations across the layer is 1.0. Can be used in the output layer of multilayer perceptrons for classification problems, so that the outputs can be interpreted as probabilities of class membership.

Source of variation A factor in a study that can be expected to increase the variability among the individuals studied.

Space plots This type of graph offers a distinctive means of representing 3D Scatterplot data through the use of a separate X-Y plane positioned at a user-selectable level of the vertical Z-axis (which "sticks up" through the middle of the plane).

The Space Plots specific layout may facilitate exploratory examination of specific types of three-dimensional data. It is recommended to assign variables to axes such that the variable that is most likely to discriminate between patterns of relation among the other two is specified as Z.

See also, Data Rotation (in 3D space).

Spatial sampling This is an area of survey sampling concerned with sampling in two (or more) dimensions. For example, sampling of fields or other planar areas.

Spearman R Spearman R can be thought of as the regular Pearson product-moment correlation coefficient (Pearson r); that is, in terms of the proportion of variability accounted for, except that Spearman R is computed from ranks. As mentioned above, Spearman R assumes that the variables under consideration were measured on at least an ordinal (rank order) scale; that is, the individual observations (cases) can be ranked into two ordered series.

Spearman rank correlation coefficient The Spearman rank correlation coefficient is one example of a correlation coefficient. It is usually calculated on occasions when it is not convenient, economic, or even possible to give actual values to variables, but only to assign a rank order to instances of each variable. It may also be a better indicator that a relationship exists between two variables when the relationship is non-linear.

Commonly used procedures, based on the Pearson's Product Moment Correlation Coefficient, for making inferences about the population correlation coefficient make the implicit assumption that the two variables are jointly normally distributed. When this assumption is not justified, a non-parametric measure such as the Spearman Rank Correlation Coefficient might be more appropriate.

See also correlation coefficient.

Specification If the strength of the association between two variables differs according the level of a third variable, the third variable specifies the relationship. Also called interaction.

Specificity For a diagnostic procedure, the conditional probability of a negative result given that disease is absent.

Spectral plot The original application of this type of plot was in the context of spectral analysis in order to investigate the behavior of non-stationary time series. On the horizontal axes one can plot the frequency of the spectrum against consecutive time intervals, and indicate on the Z-axis the spectral densities at each interval.

Spectral plots have clear advantages over the regular 3D Scatterplots when you are interested in examining how a relationship between two variables changes across the levels of a third variable.

The Spectral Plot makes it easier to see that the relationship between Pressure and Yield changes from an "inverted U" to a "U".

See also, Data Rotation (in 3D space).

Spikes (3D graphs) In this type of graph, individual values of one or more series of data are represented along the X-axis as a series of "spikes" (point symbols with lines descending to the base plane). Each series to be plotted is spaced along the Y-axis. The "height" of each spike is determined by the respective value of each series.

Spline (2D graphs) A curve is fitted to the XY coordinate data using the bicubic spline smoothing procedure.

Spline (3D graphs) A surface is fitted to the XYZ coordinate data using the bicubic spline smoothing procedure.

Split selection (for classification trees) Split selection for classification trees refers to the process of selecting the splits on the predictor variables which are used to predict membership in the classes of the dependent variable for the cases or objects in the analysis. Given the hierarchical nature of classification trees, these splits are selected one at time, starting with the split at the root node, and continuing with splits of resulting child nodes until splitting stops, and the child nodes which have not been split become terminal nodes.

Spread, measure of See inter-quartile range, range, and standard deviation.

SPSS A computer program used for the management and analysis of social science data.

Spurious A relationship between two variables is spurious if it disappears when a third variable is controlled.

Spurious correlations Correlations that are due mostly to the influences of one or more "other" variables. For example, there is a correlation between the total amount of losses in a fire and the number of firemen that were putting out the fire; however, what this correlation does not indicate is that if you call fewer firemen then you would lower the losses. There is a third variable (the initial size of the fire) that influences both the amount of losses and the number of firemen. If you "control" for this variable (e.g., consider only fires of a fixed size), then the correlation will either disappear or perhaps even change its sign. The main problem with spurious correlations is that we typically do not know what the "hidden" agent is. However, in cases when we know where to look, we can use partial correlations that control for (i.e., partial out) the influence of specified variables.

See also Correlation, partial correlation, basic statistics, multiple regression, structural equation modelling (SEPATH).

SQL (structured query language) Enables you to query an outside data source about the data it contains. You can use a SQL statement in order to specify the desired tables, fields, rows, etc. to return as data.

Square root law The Square-Root Law says that the standard error (SE) of the sample sum of n random draws with replacement from a box of tickets with numbers on them is

SE(sample sum) = $n^{1/2}$×SD(box),

And the standard error of the sample mean of n random draws with replacement from a box of tickets is

SE(sample mean) = $n^{-1/2}$×SD(box),

where SD(box) is the standard deviation of the list of the numbers on all the tickets in the box (including repeated values).

Square root of the signal to noise ratio (F) This standardized measure of effect size is used in the Analysis of Variance to characterize the overall level of population effects, and is very similar to the RMSSE. It is the square root of the sum of squared standardized effects divided by the number of effects. For example, in a 1-Way ANOVA, with J groups, f is calculated as

$$f = \sqrt{\frac{\sum_{i=1}^{n}\left(\frac{\alpha_j}{\sigma}\right)^2}{J}} = \sqrt{\frac{\sigma_\alpha^2}{\sigma^2}} = \frac{\sigma_\alpha}{\sigma}$$

Square root Used to transform the squared distance activation in an sofm network or Cluster network to the actual distance as an output.

Stacking (stacked generalization) The concept of stacking (short for Stacked Generalization) applies to the area of predictive data mining, to combine the predictions from multiple models. It is particularly useful when the types of models included in the project are very different.

Suppose your data mining project includes tree classifiers, such as C&RT and chaid, linear discriminant analysis, and Neural Networks. Each computes predicted classifications for a crossvalidation sample, from which overall goodness-of-fit statistics (e.g., misclassification rates) can be computed. Experience has shown that combining the predictions from multiple methods often yields more accurate predictions than can be derived from any one method. In stacking, the predictions from different classifiers are used as input into a meta-learner, which attempts to combine the predictions to create a final best predicted classification. So, for example, the predicted classifications from the tree classifiers, linear model, and the neural network classifier(s) can be used as input variables into a neural network meta-classifier, which will attempt to "learn" from the data how to combine the predictions from the different models to yield maximum classification accuracy.

Other methods for combining the prediction from multiple models or methods (e.g., from multiple datasets used for learning) are Boosting and Bagging (Voting).

Standard deviation Standard deviation is a measure of the spread or dispersion of a set of data.

It is calculated by taking the square root of the variance and is symbolised by s.d, or s. That is:

$$\sqrt{V(X)} = \sqrt{\sigma^2} = s$$

The more widely the values are spread out, the larger the standard deviation. For example, say we have two separate lists of exam results from a class of 30 students; one ranges from 31% to 98%, the other from 82% to 93%, then the standard deviation would be larger for the results of the first exam. The standard deviation (this term was first used by Pearson, 1894) is a commonly-used measure of variation. The standard deviation of a population of values is computed as:

$\sigma = [\Sigma (x_i-\mu)^2/N]^{1/2}$

where μ is the population mean

N is the population size.

The sample estimate of the population standard deviation is computed as:

$s = [\Sigma (x_i\text{-x-bar})^2/n\text{-}1]^{1/2}$

where xbar is the sample mean

n is the sample size.

Standard error (SE) The Standard Error of a random variable is a measure of how far it is likely to be from its expected value; that is, its scatter in repeated experiments. The SE of a random variable X is defined to be

$$SE(X) = [E((X - E(X))^2)]^{½}.$$

That is, the standard error is the square-root of the expected squared difference between the random variable and its expected value. The SE of a random variable is analogous to the SD of a list.

Standard error of the difference The standard error of a test statistic obtained from comparing two samples.

Standard error of the mean The standard error of the mean (first used by Yule, 1897) is the theoretical standard deviation of all sample means of size n drawn from a population and depends on both the population variance (sigma) and the sample size (n) as indicated below:

$$\sigma_{\bar{x}} = (\sigma^2/n)^{1/2}$$

where σ^2 is the population variance and n is the sample size.

Since the population variance is typically unknown, the best estimate for the standard error of the mean is then calculated as:

$$S_{\bar{x}} = (s^2/n)^{1/2}$$

where s^2 is the sample variance (our best estimate of the population variance) and n is the sample size.

Standard error of the proportion This is the standard deviation of the distribution of the sample proportion over repeated samples.

Standard normal curve A normal curve after standardisation to z scores.

Standard normal distribution A Normal distribution with mean 0 and standard deviation 1.

Standard programming language A programming language which has a publicly agreed common form across several different types of computer. Such standardisation allows a program to be transported conveniently between the different types of computer and is thus suitable for communicating general ideas about programming. Some standard programming languages relevant to the present context are : fortran, pascal, 'C'. There are a number of widely available programs for translating source programs from one standard programming language to another.

Standard residual value This is the standardized residual value (observed minus predicted divided by the square root of the residual mean square).

See also, Mahalanobis distance, deleted residual and Cook's distance.

Standard units A variable (a set of data) is said to be in standard units if its mean is zero and its standard deviation is one. You transform a set of data into standard units by subtracting the mean from each element of the list, and dividing the results by the standard deviation. A random variable is said to be in standard units if its expected value is zero and its standard error is one. You transform a random variable to standard units by subtracting its expected value then dividing by its standard error.

Standardization While in the everyday language, the term "standardization" means - converting to a common standard or making something conform to a standard (i.e., its meaning is similar to the term "normalization" in data analysis), in statistics, this term has a very specific meaning and refers to the transformation of data by subtracting each value from some reference value (typically a sample mean) and diving it by the standard deviation (typically a sample SD). This important transformation will bring all values (regardless of their distributions and original units of measurement) to compatible units from a distribution with a mean of 0 and a standard deviation of 1. This transformation has a wide variety of applications because it makes the distributions of values easy to compare across variables and/or subsets. If applied to the input data, standardization also makes the results of a variety of statistical techniques entirely independent of the ranges of values or the units of measurements.

Standardize To transform into standard units.

Standardized DFFITS This is another measure of impact of the respective case on the regression equation. The formula for standardized dffits is

$$SDFIT_i = DFFIT_i/(s_i(\tilde{h}_i)^{1/2})$$

where h_i is the leverage for the ith case and

$$\tilde{h}_i = 1/N + h_i$$

Standardized effect (ES) A statistical effect expressed in convenient standardized units. For example, the standardized effect in a 2 Sample t-test is the difference between the two means, divided by the standard deviation, i.e.,

$$E_s = (\mu_1 - \mu_2)/\text{ standard deviation.}$$

Standardized value Also called Z-score.

A value measured in standard deviations from the mean.

Stationary population A hypothetical population of individuals having a fixed number (typically 100,000) of births every year.

Stationary series (in time series) In Time Series analysis, a stationary series has a constant mean, variance, and autocorrelation through time (i.e., seasonal dependencies have been removed via Differencing).

Statistic A statistic is a quantity that is calculated from a sample of data. It is used to give information about unknown values in the corresponding population. For example, the average of the data in a sample is used to give information about the overall average in the population from which that sample was drawn.

It is possible to draw more than one sample from the same population and the value of a statistic will in general vary from sample to sample. For example, the average value in a sample is a statistic. The average values in more than one sample, drawn from the same population, will not necessarily be equal.

Statistics are often assigned Roman letters (e.g. m and s), whereas the equivalent unknown values in the population (parameters) are assigned Greek letters (e.g. μ and σ).

Statistica data miner StatSoft's statistica Data Miner offers the most comprehensive selection of data mining solutions on the market, with an icon-based, extremely easy-to-use user interface (optionally Web browser based via webstatistica) and a deployment engine. It features a selection of completely integrated, and automated, ready to deploy "as is" (but also easily customizable) systems of specific data mining solutions for a wide variety of business applications. A designated SPC version (QC Data Stream Analyzer) to mine/analyze large streams of QC data is also available. The product is offered optionally with deployment and on-site training services. The data mining solutions are driven by powerful procedures from five modules, which can also be used interactively, and/or used to build, test, and deploy new solutions:

General Slicer/Dicer Explorer (with optional OLAP).

Statistica enterprise wide SPC system (SEWSS) StatSoft's statistica Enterprise-wide SPC System is an enterprise-wide statistical software program is based on state-of-the-art connectivity technologies. *SEWSS* is designed for local and global enterprise quality control and improvement applications, including Six Sigma; it offers real-time monitoring and alarm notification for the production floor, a comprehensive set of analytical tools for engineers, sophisticated reporting features for management, Six Sigma reporting options, and much more. It also features:

Web-enabled user interface and reporting tools Simplistic operator interface for data entry and charting Real-time analytical tools Groupware functionality for sharing queries, special applications, etc. Wizard-driven system administration tools One-click access to analyses and reports Built-in security system Open-ended alarm notification including cause/action prompts Interactive querying facilities Integration with external applications (MicroSoft Word, MicroSoft Excel, browsers) Unlimited options for expansion and customization.

Statistica neural networks StatSoft's statistica neural networks software contains the most comprehensive selection of neural network methods with intelligent problem solvers and automatic wizards; a C-code generator add-on is available. It also features:

Intelligent Problem Solver Wizard

Automatic Search for Best Architecture

Multilayer Perceptrons

Radial Basis Function Networks, and Many Others

Self-Organizing Feature Map

Back Propogation

Conjugate Gradient Descent

Numerous Analytical Graphs Resampling (Cross Validation, Bootstrap)

Sensitivity Analysis, ROC Curves

Network Ensembles.

Statistical inference Statistical Inference makes use of information from a sample to draw conclusions (inferences) about the population from which the sample was taken.

Statistical power The probability of rejecting a false statistical null hypothesis.

Statistical process control (SPC) The term statistical process control (SPC) is typically used in context of manufacturing processes (although it may also pertain to services and other activities), and it denotes statistical methods used to monitor and improve the quality of the respective operations. By gathering information about the various stages of the process and performing statistical analysis on that information, the SPC engineer is able to take necessary action (often preventive) to ensure that the overall process stays in-control and to allow the product to meet all desired specifications. SPC involves monitoring processes, identifying problem areas, recommending methods to reduce variation and verifying that they work, optimizing the process, assessing the reliability of parts, and other analytic operations. SPC uses such basic statistical quality control methods as quality control charts (Sheward, Pareto, and others), capability analysis, gage repeatability/reproducibility analysis, and reliability analysis. However, also specialized experimental methods (DOE) and other advanced statistical techniques are often part of global SPC systems. Important components of effective, modern SPC systems are real-time access to data and facilities to document and respond to incoming QC data on-line, efficient central

QC data warehousing, and groupware facilities allowing QC engineers to share data and reports.

See also, Quality Control and Process Analysis.

Statistical significance (p-level) The statistical significance of a result is an estimated measure of the degree to which it is "true" (in the sense of "representative of the population"). More technically, the value of the p-level represents a decreasing index of the reliability of a result. The higher the p-level, the less we can believe that the observed relation between variables in the sample is a reliable indicator of the relation between the respective variables in the population. Specifically, the p-level represents the probability of error that is involved in accepting our observed result as valid, that is, as "representative of the population." For example, the p-level of .05 (i.e.,1/20) indicates that there is a 5% probability that the relation between the variables found in our sample is a "fluke." In other words, assuming that in the population there was no relation between those variables whatsoever, and we were repeating experiments like ours one after another, we could expect that approximately in every 20 replications of the experiment there would be one in which the relation between the variables in question would be equal or stronger than in ours. In many areas of research, the p-level of .05 is customarily treated as a "border-line acceptable" error level.

Statistically significant Loosely just called significant.

Indicating that a set of data provides sufficient evidence to justify rejecting a null hypothesis.

Statistics A branch of mathematics that describes and reasons from numerical observations; or descriptive measures of a sample.

Steepest descent iterations When initial values for the parameters are far from the ultimate minimum, the approximate Hessian used in the Gauss-Newton procedure may fail to yield a proper step direction during iteration. In this case, the program may iterate into a region of the parameter space from which recovery (i.e., successful iteration to the true minimum point) is not possible. One option offered by Structural Equation Modeling is to precede the Gauss-Newton procedure with a few iterations utilizing the "method of steepest descent." In the steepest descent approach, values of the parameter vector q on each iteration are obtained as

$$\hat{\theta}_{k+1} = \hat{\theta}_k + \lambda_k g_k$$

In simple terms, what this means is that the Hessian is not used to help find the direction for the next step. Instead, only the first derivative information in the gradient is used.

Stem and leaf diagram An exploratory data analytic method to show the distribution of a continuous variable

Stem and leaf plot A stem and leaf plot is a way of summarising a set of data measured on an interval scale. It is often used in exploratory data analysis to illustrate the major features of the distribution of the data in a convenient and easily drawn form.

A stem and leaf plot is similar to a histogram but is usually a more informative display for relatively small data sets (<100 data points). It provides a table as well as a picture of the data and from it we can readily write down the data in order of magnitude, which is useful for many statistical procedures.

We can compare more than one data set by the use of multiple stem and leaf plots. By using a back-to-back stem and leaf plot, we are able to compare the same characteristic in two different groups, for example, pulse rate after exercise of smokers and non-smokers.

Stemming An important pre-processing step before indexing input documents for text mining is the stemming of words. The term stemming refers to the reduction of words to their roots so that, for example, different grammatical forms or declinations of verbs are identified and indexed (counted) as the same word. For example, stemming will ensure that both "travel" and "traveled" will be recognized by the program as the same word.

Steps Repetitions of a particular analytic or computational operation or procedure. For example in the neural network time series analysis, the number of consecutive time steps from which input variable values should be drawn to be fed into the neural network input units.

Stepwise regression A model-building technique which finds subsets of predictor variables that most adequately predict responses on a dependent variable by linear (or nonlinear) regression, given the specified criteria for adequacy of model fit.

Stepwise selection A procedure for exploratory analysis that alternates using forward selection and backward elimination in order to find the most satisfactory set of variables to use in a regression equation.

Stevens' typology This is widely-observed scheme of distinctions between types of measurement scales according to the meaningfulness of arithmetic which may be performed upon data values. The types are : nominal scale versus ordinal scale versus interval scale versus ratio scale.

Stiffness parameter (in fitting options) The function that controls the weight is determined by the Stiffness parameter which can be modified. Thus, the stiffness parameter determines the degree to which the fitted curve depends on local configurations of the analyzed values.

The lower the coefficient, the more the shape of the curve is influenced by individual data points (i.e., the curve "bends" more to accommodate individual values and subsets of values).

The range of the stiffness parameters is 0 < s < 1. Large values of the parameter produce smoother curves that adequately represent the overall pattern in the data set at the expense of local details.

Stochastically larger One subpopulation (or subsample) is said to be stochastically larger than another (with respect to a variable) if every percentile of the first is greater than or equal to the same percentile of the second. The cdf of a stochastially larger subpopulation will lie to the right (and below) the cdf of the stochastically smaller one.

Stopping conditions The iterative gradient-descent training algorithms (back propagation, Quasi-Newton, conjugate gradient descent, Levenberg-Marquardt, quick propagation, Delta-bar-Delta, and Kohonen) all attempt to reduce the training error on each epoch.

You specify a maximum number of epochs for these iterative algorithms. However, you can also define stopping conditions that may cause training to determine earlier.

Specifically, training may be stopped when:

the error drops below a given level;

the error fails to improve by a given amount over a given number of epochs.

The conditions are cumulative; i.e., if several stopping conditions are specified, training ceases when any one of them is satisfied. In particular, a maximum number of epochs must always be specified.

The error-based stopping conditions can also be specified independently for the error on the training set and the error on the selection set (if any).

Stopping conditions During an iterative process (e.g., fitting, searching, training), the conditions which must be true for the process to stop. (For example, in neural networks, the stopping conditions include the maximum number of epochs, target error performance and the minimum error improvement thresholds.

Stopping rule (in classification trees) The stopping rule for a classification tree refers to the criteria that are used for determining the "right-sized" classification tree, that is, a classification tree with an appropriate number of splits and optimal predictive accuracy.

Stratified This is a feature of an experimental design whereby a scheme of observations is repeated entirely using further sets (strata) of experimental units, with each such further set distinguished by a level of a categorical variable which is distinct from any categorical variables used to define the experimnatl design within a single set (stratum). The data from the various strata are regarded as distinct. This situation occurs when attempting to make inferences based upon the results of several similar independent experiments.

Stratified random sampling In general, random sampling is the process of randomly selecting observations from a population, to create a subsample that "represents" the observations in that population. In stratified sampling one usually applies specific (identical or different) sampling fractions to different groups (strata) in the population to draw the sample.

Stratified sample A sample obtained by combining simple random samples taken from several disjoint subpopulations or strata.

Stratified sampling There may often be factors which divide up the population into sub-populations (groups / strata) and we may expect the measurement of interest to vary among the different sub-populations. This has to be accounted for when we select a sample from the population in order that we obtain a sample that is representative of the population. This is achieved by stratified sampling.

A stratified sample is obtained by taking samples from each stratum or sub-group of a population.

When we sample a population with several strata, we generally require that the proportion of each stratum in the sample should be the same as in the population.

Stratified sampling techniques are generally used when the population is heterogeneous, or dissimilar, where certain homogeneous, or similar, sub-populations can be isolated (strata). Simple random sampling is most appropriate when the entire population from which the sample is taken is homogeneous. Some reasons for using stratified sampling over simple random sampling are :

a) the cost per observation in the survey may be reduced;

b) estimates of the population parameters may be wanted for each sub-population;

c) increased accuracy at given cost.

Example

Suppose a farmer wishes to work out the average milk yield of each cow type in his herd which consists of Ayrshire, Friesian, Galloway and Jersey cows. He could divide up his herd into the four sub-groups and take samples from these.

Stratum In random sampling, sometimes the sample is drawn separately from different disjoint subsets of the population. Each such subset is called a *stratum*. (The plural of *stratum* is *strata*.) Samples drawn in such a way are called stratified samples. Estimators based on stratified random samples can have smaller sampling errors than estimators computed from simple random samples of the same size, if the average variability of the variable of interest within strata is smaller than it is across the entire population; that is, if stratum membership is associated with the variable.

For example, to determine average home prices in the U.S., it would be advantageous to *stratify* on geography, because average home prices vary enormously with location. We might divide the country into states, then divide each state into urban, suburban, and rural areas; then draw random samples separately from each such division.

Stub and Banner tables (Banner tables) Stub-and-banner tables are essentially two-way tables, except that two lists of categorical variables (instead of just two individual variables) are crosstabulated. In the Stub-and-banner table, one list will be tabulated in the columns (horizontally) and the second list will be tabulated in the rows (vertically) of the Scrollsheet.

Stub The left-hand column of a table, in which the labels describing each row are placed.

Studentized deleted residuals In addition to standardized residuals several methods (including studentized residuals, studentized deleted residuals, DFFITS, and standardized DFFITS) are available for detecting outlying values (observations with extreme values on the set of predictor variables or the dependent variable). The formula for studentized deleted residuals is given by

$SDRESID_i = DRESID_i / s_{(i)}$ for $DRESID = e_i/(1-\hbar_i)$ and where

$s_{(i)} = 1/(C-p-1)^{1/2} * ((C-p)s^2/1-h_i) - DRESID_i^2)^{1/2}$

e_i is the error for the ith case

h_i is the leverage for the ith case

p is the number of coefficients in the model

an $\hbar_i = 1/N + h_i$

Studentized residuals In addition to standardized residuals several methods (including studentized residuals, studentized deleted residuals, DFFITS, and standardized DFFITS) are available for detecting outlying values (observations with extreme values on the set of predictor variables or the dependent variable). The formula for studentized residuals is

$SRES_i = (e_i/s)/(1-_i)^{1/2}$

where e_i is the error for the ith case

h_i is the leverage for the ith case

and $\hbar_i = 1/N + h_i$

Studentized score The observed value of a statistic, minus the expected value of the statistic, divided by the estimated standard error of the statistic.

Student's *t* curve Student's *t* curve is a family of curves indexed by a parameter called the *degrees of freedom*, which can take the values 1, 2, . . . Student's *t* curve is used to approximate some probability histograms. Consider a population of numbers that are nearly normally distributed and have population mean is μ. Consider drawing a random sample of size *n* with replacement from the population, and computing the sample mean *M* and the sample standard deviation *S*. Define the random variable $T = (M - \mu)/(S/n^{1/2})$.

If the sample size *n* is large, the probability histogram of *T* can be approximated accurately by the normal curve. However, for small and intermediate values of *n*, Student's *t* curve with *n* - 1 *degrees of freedom* gives a better approximation. That is,

$P(a < T < b)$ is approximately the area under Student's T curve with $n - 1$ degrees of freedom, from a to b.

Student's t curve can be used to test hypotheses about the population mean and construct confidence intervals for the population mean, when the population distribution is known to be nearly normally distributed.

Student's T distribution The Student's t distribution has density function (for $\nu = 1, 2, ...$):

$$f(x) = \frac{\Gamma\left(\frac{\nu+1}{2}\right)}{\Gamma\left(\frac{\nu}{2}\right)} \frac{1}{\sqrt{\nu\pi}} \left(1 + \frac{x^2}{\nu}\right)^{-\frac{\nu+1}{2}}$$

where ν is the degrees of freedom

Γ (gamma) is the Gamma function

π is the constant Pi (3.14...).

Subject See individual.

Subject, experimental subject A member of the control group or the treatment group.

Subjective An interpretation of probability that maintains that a probability quantifies an aspect of one's mental state. cf. objective.

Subjective probability A subjective probability describes an individual's personal judgement about how likely a particular event is to occur. It is not based on any precise computation but is often a reasonable assessment by a knowledgeable person.

Like all probabilities, a subjective probability is conventionally expressed on a scale from 0 to 1; a rare event has a subjective probability close to 0, a very common event has a subjective probability close to 1.

A person's subjective probability of an event describes his/her degree of belief in the event.

Example

A Rangers supporter might say 'I believe that Rangers have probability 0.9 of winning the Scottish Premier Division this year since they have been playing really well'.

Subsample A subsample is a selection of cases from a sample. For example, if you remove all the males from a sample of both males and females, you are left with a subsample of females.

Subset A subset of a given set is a collection of things that belong to the original set. Every element of the subset must belong to the original set, but not every element of the original set need be in a subset (otherwise, a subset would always be identical to the set it came from).

Success The value of a binary variable that is identified as the event of interest, often coded as 1.

Sum of squares A component of an analysis of variance table, usually computed by summing the squares of data values after combinations of averaging and differencing.

Sum squared The error is the sum of the squared differences between the target and actual output values on each output unit. This is the standard error function used in regression problems. It can also be used for classification problems, giving robust performance in estimating discriminant functions, although arguably entropy functions are more appropriate for classification, as they correspond to maximum likelihood decision making (on the assumption that the generating distribution is drawn from the exponential family), and allow outputs to be interpreted as probabilities.

Sum squared error function Levenberg-Marquardt is only defined for the sum squared error function. If you select a different error function for your network, it will be ignored during Levenberg-Marquardt training. It is usually therefore only appropriate for regression networks.

Note: Like other iterative algorithms, Levenberg-Marquardt does not train radial units. Therefore, you can use it to optimize the non-radial layers of radial basis function networks even if there are a large number of weights in the radial layer, as those are ignored by Levenberg-Marquardt. This is significant as it is typically the radial layer that is very large in such networks.

Levenberg-Marquardt works by making the assumption that the underlying function being modeled by the neural network is linear. Based on this calculation, the minimum can be determined exactly in a single step. The calculated minimum is tested, and if the error there is lower, the algorithm moves the weights to the new point. This process is repeated iteratively on each generation. Since the linear assumption is ill-founded, it can easily lead Levenberg-Marquardt to test a point that is inferior (perhaps even wildly inferior) to the current one. The clever aspect of Levenberg-Marquardt is that the determination of the new point is actually a compromise between a step in the direction of steepest descent and the above-mentioned leap. Successful steps are accepted and lead to a strengthening of the linearity assumption (which is approximately true near to a minimum). Unsuccessful steps are rejected and lead to a more cautious downhill step. Thus, Levenberg-Marquardt continuously switches its approach and can make very rapid progress.

Supervised and unsupervised learning An important distinction in machine learning, and also applicable to data mining, is that between supervised and unsupervised learning algorithms. The term "supervised" learning is usually applied to cases in which a particular classification is already observed and recorded in a training sample, and you want to build a model to predict those classifications (in a new testing sample). For example, you may have a data set that contains information about who from among a list of customers targeted for a special promotion responded to that offer. The purpose of the classification analysis would be to build a model to predict who (from a different list of new potential customers) is likely to respond to the same (or a similar) offer in the future. You may want to review the methods discussed in general classification and regression trees (GC&RT), general chaid models (GCHAID), Discriminant Function Analysis and general discriminant analysis (GDA), MARSplines (multivariate adaptive regression splines), and neural networks to learn about different techniques that can be used to build or fit models to data where the outcome variable of interest (e.g., customer did or did not respond to an offer) was observed. These methods are called supervised learning algorithms because the learning (fitting of models) is "guided" or "supervised" by the observed classifications recorded in the data file.

In unsupervised learning, the situation is different. Here the outcome variable of interest is not (and perhaps cannot be) directly observed. Instead, we want to detect some "structure" or clusters in the data that may not be trivially observable. For example, you may have a database of customers with various demographic indicators and variables potentially relevant to future purchasing behavior. Your goal would be to find market segments, i.e., groups of observations that are relatively similar to each other on certain variables; once identified, you could then determine how best to reach one or more clusters by providing certain goods or services you think may have some special utility or appeal to individuals in that segment (cluster). This type of task calls for an unsupervised learning algorithm, because learning (fitting of models) in this case cannot be guided by previously known classifications. Only after identifying certain clusters can you begin to assign labels, for example, based on subsequent research (e.g., after identifying one group of customers as "young risk takers").

There are several methods available for unsupervised learning, including Principal Components and Classification Analysis, Factor Analysis, Multidimensional Scaling, Correspondence Analysis, Neural Networks, Self-Organizing Feature Maps (SOFM, Kohonen networks); particularly powerful algorithms for pattern recognition and clustering are the EM and k-Means clustering algorithms.

Support value (association rules) When applying (in data or text mining) algorithms for deriving association rules of the general form If Body then Head (e.g., If (Car=Porsche and Age<20) then (Risk=High and Insurance=High)), the Support value is computed as the joint probability (relative frequency of co-occurrence) of the Body and Head of each association rule.

Support vector A set of points in the feature space that determines the boundary between objects of different class memberships.

Support vector machine (SVM) A classification method based on the maximum margin hyperplane.

Suppressor variable A suppressor variable (in Multiple Regression) has zero (or close to zero) correlation with the criterion but is correlated with one or more of the predictor variables, and therefore, it will suppress irrelevant variance of independent variables. For example, you are trying to predict the times of runners in a 40 meter dash. Your predictors are Height and Weight of the runner. Now, assume that Height is not correlated with Time, but Weight is. Also assume that Weight and Height are correlated. If Height is a suppressor variable, then it will suppress, or control for, irrelevant variance (i.e., variance that is shared with the predictor and not the criterion), thus increasing the partial correlation. This can be viewed as ridding the analysis of noise.

Let t = Time, h = Height, w - Weight, r_{th} = 0.0, r_{tw} = 0.5, and r_{hw} = 0.6.

Weight in this instance accounts for 25% (R_{tw}**2 = 0.5**2) of the variability of Time. However, if Height is included in the model, then an additional 14% of the variability of Time is accounted for even though Height is not correlated with Time:

$R_{t.hw}$**2 = 0.5**2/(1 – 0.6**2) = 0.39

Sure event An event that always happens commonly denoted by ∧. (This is a capital letter "Omega" - some browsers may not display it.)

Surface plot (from raw data) This sequential plot fits a spline-smoothed surface to each data point. Successive values of each series are plotted along the X-axis, with each successive series represented along the Y-axis.

Survey See sample survey.

Survival analysis Survival analysis (exploratory and hypothesis testing) techniques include descriptive methods for estimating the distribution of survival times from a sample, methods for comparing survival in two or more groups, and techniques for fitting linear or non-linear regression models to survival data. A defining characteristic of survival time data is that they usually include so-called censored observations, e.g., observations that "survived" to a certain point in time, and then dropped out from the study (e.g., patients who are discharged from a hospital). Instead of discarding such observations from the data analysis all together (i.e., unnecessarily loose potentially useful information) survival analysis techniques can accommodate censored observations, and "use" them in statistical significance testing and model fitting.

Typical survival analysis methods include life table, survival distribution, and Kaplan-Meier survival function estimation, and additional techniques for comparing the survival in two or more groups. Finally, Survival analysis includes the use of regression models for estimating the relationship of (multiple) continuous variables to survival times.

Survivor function The function relating a length of time to the probability that an individual survives (at least) that time.

Survivorship function The survivorship function (commonly denoted as R(t)) is the complement to the cumulative distribution function (i.e., R(t)=1-F(t)); the survivorship function is also referred to as the reliability or survival function (since it describes the probability of not failing or of surviving until a certain time t).

Sweeping The sweeping transformation of matrices is commonly used to efficiently perform stepwise multiple regression or similar analyses; a modified version of this transformation is also used to compute the g2 generalized inverse. The forward sweeping transformation for a column k can be summarized in the following four steps (where the e's refer to the elements of a symmetric matrix):

$e_{ij} = e_{ij} - e_{jk} * e_{kj} / e_{kk}$ for $i<>k, j<>k$

$e_{kj} = e_{kj} / e_{kk}$

$e_{ik} = e_{ik} / e_{kk}$

$e_{kk} = -1 / e_{kk}$

The reverse sweeping operation reverses the changes effected by these transformations. The sweeping operator is used extensively in General Linear Models, Multiple Regression, and similar techniques.

Symmetric distribution The probability distribution of a random variable X is symmetric if there is a number *a* such that the chance that X>=*a*+*b* is the same as the chance that X<=*a*-*b* for every value of *b*. A list of numbers has a symmetric distribution if there is a number *a* such that the fraction of numbers in the list that are greater than or equal to *a*+*b* is the same as the fraction of numbers in the list that are less than or equal to *a*-*b*, for every value of *b*. In either case, the histogram or the probability histogram will be symmetrical about a vertical line drawn at *x*=*a*.

Symmetric matrix A matrix is symmetric if the transpose of the matrix is itself (i.e., A = A'). In other words, the lower triangle of the square matrix is a "mirror image" of the upper triangle with 1's on the diagonal.

|1 2 3 4|

|2 1 5 6|

|3 5 1 7|

|4 6 7 1|

Symmetrical distribution If you split the distribution in half at its mean (or median), then the distribution of values would be a "mirror image" about this central point.

Symmetry Symmetry is implied when data values are distributed in the same way above and below the middle of the sample.

Symmetrical data sets

a) are easily interpreted;

b) allow a balanced attitude to outliers, that is, those above and below the middle value (median) can be considered by the same criteria;

c) allow comparisons of spread or dispersion with similar data sets.

Many standard statistical techniques are appropriate only for a symmetric distributional form. For this reason, attempts are often made to transform skew-symmetric data so that they become roughly symmetric.

Systematic error An error that affects all the measurements similarly. For example, if a ruler is too short, everything measured with it will appear to be longer than it really is (ignoring random error). If your watch runs fast, every time interval you measure with it will appear to be longer than it really is (again, ignoring random error). Systematic errors do not tend to average out.

Systematic random sample A systematic sample starting at a random point in the listing of units in the of frame, instead of starting at the first unit. Systematic random sampling is better than systematic sampling, but typically not as good as simple random sampling.

Systematic sample A systematic sample from a frame of units is one drawn by listing the units and selecting every kth element of the list. For example, if there are N units in the frame, and we want a sample of size $N/10$, we would take every tenth unit: the first unit, the eleventh unit, the 21st unit, *etc.* Systematic samples are not random samples, but they often behave essentially as if they were random, if the order in which the units appears in the list is haphazard. Systematic samples are a special case of cluster samples.

T

T distribution A probability distribution often used when the population standard deviation is not known or when the sample size is small.

T test An hypothesis test based on approximating the probability histogram of the test statistic by Student's *t* curve. *t* tests usually are used to test hypotheses about the mean of a population when the sample size is intermediate and the distribution of the population is known to be nearly normal.

Tabled value The value of a computed statistic used as a threshold to decide whether the null hypothesis will be rejected.

Tail An area at the extreme of a randomisation distribution, where the degree of extremity is sufficient to be notable judged against some nominal alpha criterion value. Also See branch-and bound, re-randomisation test, tail definition policy.

Tail definition policy This is a defined method for dividing a discrete distribution into a tail area and a body area. The scope for differing policies arises due to the non-infinitesmal amount of probability measure which may be associated with the actual outome value. The conventional policy, based upon considerations of simplicity and of conservatism in terms of alpha, is to include the whole of the weight of outcomes equal to the actual outcome as part of the tail.

Target error You can specify a target error level, for the training subset, the selection subset, or both. If the RMS falls below this level, training ceases.

Target population The target population is the entire group a researcher is interested in; the group about which the researcher wishes to draw conclusions.

Example

Suppose we take a group of men aged 35-40 who have suffered an initial heart attack. The purpose of this study could be to compare the effectiveness of two drug regimes for delaying or preventing further attacks. The target population here would be all men meeting the same general conditions as those actually included in the study.

Test statistic A statistic measuring the strength of the pattern which a statistical test undertakes to detect. In the context of re-randomisation tests one is concerned with the distribution of the values of the test statistic over the randomisation set. An example of a test statistic is the difference of means as employed in the pitman permutation test. A statistic that is used to identify extreme events, and hence to decide whether to reject the null hypothesis. Typically, larger values of the test statistics correspond to more extreme events, although sometimes the setup is such that one rejects for small values of the test statistic. A statistic used to test hypotheses. An hypothesis test can be constructed by deciding to reject the null hypothesis when the value of the test statistic is in some range or collection of ranges. To get a test with a specified significance level, the chance when the null hypothesis is true that the test statistic falls in the range where the hypothesis would be rejected must be at most the specified significance level. The Z statistic is a common test statistic.

Tied ranks In a nonparametric test involving ranked data, if two data have tied values then they will deserve to receive the same rank value. It is generally agreed that this should be the average of the ranks which would have been assigned if the values had been discernably unequal. Thus, the ranks assigned to a set of 6 data, with ties present might emerge as sets such as : 1,3,3,3,5,6 or 1,2,3.5,3.5,5,6. The possibility of tied ranks leads to elaborations in the otherwise-standard tasks of computing or tabulating randomisation distributions where data are replaced by ranks.

Tied values Where data are represented by ranks, tied values lead to tied ranks. Whether or not data are rep[resnted by ranks, for any test statistic the occurrence of tied values will increase the extent to which a randomisation distribution will be a discrete distribution rather than a continuous distribution.

Time series A time series is a sequence of observations which are ordered in time (or space). If observations are made on some phenomenon throughout time, it is most sensible to display the data in the order in which they arose, particularly since successive observations will probably be dependent. Time series are best displayed in a scatter plot. The series value X is plotted on the vertical axis and time t on the horizontal axis. Time is called the independent variable (in this case however, something over which you have little control).

There are two kinds of time series data:

1. Continuous, where we have an observation at every instant of time e.g. lie detectors, electrocardiograms. We denote this using observation X at time t, X(t).

2. Discrete, where we have an observation at (usually regularly) spaced intervals. We denote this as Xt.

Examples

1. Economics: weekly share prices; monthly profits.

2. Meteorology: daily rainfall; wind speed; temperature.

3. Sociology: crime figures (number of arrests, etc); employment figures.

Total Number of cases of that class.

Transformation A mathematical procedure employed to change the scale of a variable from one set of values into another set of values, without changing their relative order. This procedure is often employed in order to produce a more linear relationship between two continuous variables. Examples include transforming proportions into percentages by multiplying each value by 100 and, in regression analysis, transforming raw scores into logarithmic scores. Transformations turn lists into other lists, or variables into other variables. For example, to transform a list of temperatures in degrees Celsius into the corresponding list of temperatures in degrees Fahrenheit, you multiply each element by 9/5, and add 32 to each product. This is an example of an affine transformation: multiply by something and add something ($y = ax + b$ is the general affine transformation of x; it's the familiar equation of a straight line). In a linear transformation, you only multiply by something ($y = ax$). Affine transformations are used to put variables in standard units. In that case, you subtract the mean and divide the results by the SD. This is equivalent to multiplying by the reciprocal of the SD and adding the negative of the mean, divided by the SD, so it is an affine transformation. Affine transformations with positive multiplicative constants have a simple effect on the mean, median, mode, quartiles, and other percentiles: the new value of any of these is the old one, transformed using exactly the same formula. When the multiplicative constant is negative, the mean, median, mode, are still transformed by the same rule, but quartiles and percentiles are reversed: the qth quantile of the transformed distribution is the transformed value of the 1-qth quantile of the original distribution (ignoring the effect of data spacing). The effect of an affine transformation on the SD, range, and IQR, is to make the new value the old value times the absolute value of the number you multiplied the first list by: what you added does not affect them.

Transformation to linearity Transformations allow us to change all the values of a variable by using some mathematical operation, for example, we can change a number, group of numbers, or an equation by multiplying or dividing by a constant or taking the square root. A transformation to linearity is a transformation of a response variable, or independent variable, or both, which produces an approximate linear relationship between the variables.

Transformation to normality If there is evidence of marked non-normality then we may be able to remedy this by applying suitable transformations.

The more commonly used transformations which are appropriate for data which are skewed to the right with increasing strength (positive skew) are:

$\frac{1}{x}$, log x and $\sqrt{x}$

The more commonly used transformations which are appropriate for data which are skewed to the left with increasing strength (negative skew) are:

x^2, x^3 and e^x

Where the x's are the data values.

Treatment The substance or procedure studied in an experiment or observational study. At issue is whether the treatment has an effect on the outcome or variable of interest. In experiments, a treatment is something that researchers administer to experimantal units . For example, a corn field is divided into four, each part is 'treated' with a different fertiliser to see which produces the most corn; a teacher practices different teaching methods on different groups in her class to see which yields the best results; a doctor treats a patient with a skin condition with different creams to see which is most effective.

Treatments are administered to experimental units by 'level', where level implies amount or magnitude. For example, if the experimental units were given 5mg, 10mg, 15mg of a medication, those amounts would be three levels of the treatment. 'Level' is also used for categorical variables, such as Drugs A, B, and C, where the three are different kinds of drug, not different amounts of the same thing.

Treatment effect The effect of the treatment on the variable of interest. Establishing whether the treatment has an effect is the point of an experiment.

Treatment group The individuals who receive the treatment, as opposed to those in the control group, who do not.

Trend component We want to increase our understanding of a time series by picking out its main features. One of these main features is the trend component. Descriptive techniques may be extended to forecast (predict) future values.

Trend is a long term movement in a time series. It is the underlying direction (an upward or downward tendency) and rate of change in a time series, when allowance has been made for the other components.

A simple way of detecting trend in seasonal data is to take averages over a certain period. If these averages change with time we can say that there is evidence of a trend in the series. There are also more formal tests to enable detection of trend in time series.

It can be helpful to model trend using straight lines, polynomials etc.

See also time series.

See also cyclical component.

See also seasonal component.

See also irregular component.

Two by two table (2×2 table) A contingency table with two rows and two columns.

Two one sided tests (TOST) A statistical procedure for testing practical equivalence, in which the alternative hypothesis is that the parameter value lies inside an interval of acceptability, and the null hypothesis is that the parameter either exceeds the upper limit of acceptability or falls below the lower limit of acceptability.

Two sample T test A two sample t-test is a hypothesis test for answering questions about the mean where the data are collected from two random samples of independent observations, each from an underlying normal distribution:

$N(\mu_i, \sigma_i^2)$, = where $i = 12$

When carrying out a two sample t-test, it is usual to assume that the variances for the two populations are equal, that is:

$$\sigma_1^2 = \sigma_2^2 = \sigma^2$$

The null hypothesis for the two sample t-test is:

$$H_0 : \mu 1 = 2$$

That is, the two samples have both been drawn from the same population.

This null hypothesis is tested against one of the following alternative hypotheses, depending on the question posed.

$$H_1 : \mu_1 \neq \mu_2$$

$$H_1 : \mu_1 > \mu_2$$

$$H_1 : \mu_1 < \mu_2$$

Two sample test A test comparing two statistics derived from the characteristics of different samples

Two sided alternative An alternative hypothesis that asserts that a parameter value could be on either side of the value(s) asserted by the null hypothesis. Such an alternative hypothesis will say that the parameter is either greater than or less than a hypothesized value.

Two sided hypothesis test An hypothesis test of the null hypothesis that the value of a parameter, μ, is equal to a null value, μ_0, designed to have power against the alternative hypothesis that either $\mu < \mu_0$ or $\mu > \mu_0$ (the alternative hypothesis contains values on both sides of the null value). For example, a significance level 5%, two-sided *z* test of the null hypothesis that the mean of a population equals zero against the alternative that it is greater than zero would reject the null hypothesis for values of

|(sample mean)|

$|z| = \frac{\text{— — — — — — — —}}{\text{SE(sample mean)}} > 1.96.$

Two sided test A two-sided test is a statistical hypothesis test in which the values for which we can reject the null hypothesis, H0 are located in both tails of the probability distribution.

In other words, the critical region for a two-sided test is the set of values less than a first critical value of the test and the set of values greater than a second critical value of the test.

A two-sided test is also referred to as a two-tailed test of significance.

The choice between a one-sided test and a two-sided test is determined by the purpose of the investigation or prior reasons for using a one-sided test.

Example

Suppose we wanted to test a manufacturers claim that there are, on average, 50 matches in a box. We could set up the following hypotheses.

$H_0 : \mu = 50$ against $H_1 : \mu < 50$ or $H_1 : \mu > 50$

Either of these two alternative hypotheses would lead to a one-sided test. Presumably, we would want to test the null against the first alternative hypothesis since it would be useful to know if there is likely to be less than 50 matches, on average, in a box (no one would complain if they get the correct number of matches in a box or more).

Yet another alternative hypothesis could be tested against the same null, leading this time to a two-sided test:

$H_0 : \mu = 50$ against $H_1 : \mu \neq 50$

That is, nothing specific can be said about the average number of matches in a box; only that, if we could reject the null hypothesis in our test, we would know that the average number of matches in a box is likely to be less than or greater than 50.

Two tailed test A hypothesis test in which the rejection region consists of both large values and small values of the test statistic. Such a test is appropriate against two-sided alternatives. (Note, however, that F or chi-squared tests against two-sided alternatives are one-tailed.)

Two way analysis of variance Two Way Analysis of Variance is a way of studying the effects of two factors separately (their main effects) and (sometimes) together (their interaction effect).

See also factor.

See also main effect.

See also interaction.

Two way table A representation of suitable data in a table organised as rows and columns, such that the rows represent one scheme of alternatives covering the whole of the the data represented, the columns represent a further scheme of alternatives covering the whole of the data represented, and the entries in the two-way table are the counts of numbers of observations conforming to the respective cells of the two-way classification.

Type I and type II errors These refer to hypothesis testing. A Type I error occurs when the null hypothesis is rejected erroneously when it is in fact true. A Type II error occurs if the null hypothesis is not rejected when it is in fact false. See also significance level and power.

Type I error In a hypothesis test, a type I error occurs when the null hypothesis is rejected when it is in fact true; that is, H0 is wrongly rejected. For example, in a clinical trial of a new drug, the null hypothesis might be that the new drug is no better, on average, than the current drug; that is H0: there is no difference between the two drugs on average. A type I error would occur if we concluded that the two drugs produced different effects when in fact there was no difference between them.

The following table gives a summary of possible results of any hypothesis test:

		Decision	
		Reject H0	Don't reject H0
Truth	H0	Type I Error	Right Decision
	H1	Right Decision	Type II Error

A type I error is often considered to be more serious, and therefore more important to avoid, than a type II error. The hypothesis test procedure is therefore adjusted so that there is a guaranteed 'low' probability of rejecting the null hypothesis wrongly; this probability is never 0. This probability of a type I error can be precisely computed as,

Type II error In a hypothesis test, a type II error occurs when the null hypothesis H0, is not rejected when it is in fact false. For example, in a clinical trial of a new drug, the null hypothesis might be that the new drug is no better, on average, than the current drug; that is H0: there is no difference between the two drugs on average.A type II error would occur if it was concluded that the two drugs produced the same effect, that is, there is no difference between the two drugs on average, when in fact they produced different ones.

A type II error is frequently due to sample sizes being too small.

U

Unbiased Not biased; having zero bias.

Unbiased estimator An estimator whose expectation is the true parameter.

Uncontrolled experiment An experiment in which there is no control group; *i.e.*, in which the method of comparison is not used: the experimenter decides who gets the treatment, but the outcome of the treated group is not compared with the outcome of a control group that does not receive treatment.

Uncorrelated A set of bivariate data is uncorrelated if its correlation coefficient is zero. Two random variables are uncorrelated if the expected value of their product equals the product of their expected values. If two random variables are independent, they are uncorrelated. (The converse is not true in general.)

Uncountable A set is uncountable if it is not countable.

Unexplained variance In regression, unexplained variance is the variance after the variance 'explained' by the regression is subtracted from the total variance of the dependent variable. Often called the variance of the residuals.

Uniform distribution Uniform distributions model (some) continuous random variables and (some) discrete random variables. The values of a uniform random variable are uniformly distributed over an interval. For example, if buses arrive at a given bus stop every 15 minutes, and you arrive at the bus stop at a random time, the time you wait for the next bus to arrive could be described by a uniform distribution over the interval from 0 to 15.

A discrete random variable X is said to follow a Uniform distribution with parameters a and b, written X ~ Un(a,b), if it has probability distribution:

$$P(X = x) = \frac{1}{b - a}$$

where:

x = 1, 2, 3, ..., n.

A discrete uniform distribution has equal probability at each of its n values.

A continuous random variable X is said to follow a Uniform distribution with parameters a and b, written X ~ Un(a,b), if its probability density function is constant within a finite interval [a,b], and zero outside this interval, where $a \leq x \leq b$

The Uniform distribution has expected value:

$$E(X) = \frac{(a + b)}{2}$$

and variance:

$$V(X) = \frac{1}{12}(b - a)^2$$

Unimodal Having exactly one mode.

Union The union of two or more sets is the set of objects contained by at least one of the sets. The union of the events A and B is denoted "A+B", "A or B", and "AUB".

Unit A member of a population.

Unit sum Normalizes the outputs to sum to 1.0. Used in PNNs to allow the outputs to be interpreted as probabilities.

Univariate Having or having to do with a single variable. Some univariate techniques and statistics include the histogram, IQR, mean, median, percentiles, quantiles, and SD.

Univariate statistics Univariate statistics are statistics that describe the characteristics of a single variable.

Universe The set of all those to whom the research hypothesis is assumed to apply (a synonym for population)

Unknown Number of cases that could not be positively classified.

Upper confidence limit The upper endpoint of a confidence interval. This term is typically used when the lower endpoint if minus infinity.

Upper fence A value equal to the upper quartile plus 1.5×IQR.

Upper quartile (Q3) A value that exceeds the values for no more than three quarters of the individuals and is exceeded by the values of no more than one quarter of the individuals.

V

V fold cross validation In v-fold cross-validation, repeated (v) random samples are drawn from the data for the analysis, and the respective model or prediction method, etc. is then applied to compute predicted values, classifications, etc. Typically, summary indices of the accuracy of the prediction are computed over the v replications; thus, this technique allows the analyst to evaluate the overall accuracy of the respective prediction model or method in repeatedly drawn random samples. This method is customarily used in tree classification and regression.

Validity Validity concerns how well an indicator measures the concept it is designed to measure. For instance, if we wanted to measure height and all we had was a set of bathroom scales, we would not have a valid indicator since the scales measure weight. However, the scales are a very reliable measure since we get the same weight each time. However, if we used an elastic tape measure to measure height, we may get a different height each time we use it - it is very unreliable but it *is* a valid indicator since it does measure height. See reliability.

Value A measurement or classification of a variable.

Value label Value labels in spss are an optional way of explaining the numeric codes ascribed to each response of a variable. For example, code 1, indicating those respondents who are married, may be given the value label 'Married'.

Variable A numerical value or a characteristic that can differ from individual to individual. See also categorical variable, qualitative variable, quantitative variable, discrete variable, continuous variable, and random variable.

Variable centred A type of data analysis concerned with exploring relationships between variables, rather than exploring relationships between cases.

Variable label A label that may be created in spss to describe each variable. Variable labels, with a limit of 255 characters, allow you to expand on the 8-character variable name. The variable label for the variable marstat could be "Respondent's marital status".

Variable name Required by spss to identify each variable. Variable names may be up to 8 characters long and must be unique in any one data file. For example, marital status may be given the variable name marstat.

Variance The (population) variance of a random variable is a non-negative number which gives an idea of how widely spread the values of the random variable are likely to be; the larger the variance, the more scattered the observations on average.

Stating the variance gives an impression of how closely concentrated round the expected value the distribution is; it is a measure of the 'spread' of a distribution about its average value.

Variance is symbolised by V(X) or Var(X) or σ^2

The variance of the random variable X is defined to be:

$$V(X) = \sigma^2 = E[X - E(X)]^2 = E(X)^2 - [E(X)]^2$$

Where E(X) is the expected value of the random variable X.

Notes

a) the larger the variance, the further that individual values of the random variable (observations) tend to be from the mean, on average;

b) the smaller the variance, the closer that individual values of the random variable (observations) tend to be to the mean, on average;

c) taking the square root of the variance gives the standard deviation. That is:

$$\sqrt{V(X)} = \sqrt{\sigma^2} = \sigma = s$$

d) the variance and standard deviation of a random variable are always non-negative.

See also sample variance.

Variance components (in mixed model ANOVA) The term variance components is used in the context of experimental designs with random effects, to denote the estimate of the (amount of) variance that can be attributed to those effects. For example, if one were interested in the effect that the quality of different schools has on academic proficiency, one could select a sample of schools to estimate the amount of variance in academic proficiency (component of variance) that is attributable to differences between schools.

Variance inflation factor (VIF) The diagonal elements of the inverse correlation matrix (i.e., -1 times the diagonal elements of the sweep matrix) for variables that are in the equation are also sometimes called variance inflation factors. This terminology denotes the fact that the variances of the standardized regression coefficients can be computed as the product of the residual variance (for the correlation transformed model) times the respective diagonal elements of the inverse correlation matrix. If the predictor variables are uncorrelated, then the diagonal elements of the inverse correlation matrix are equal to 1.0; thus, for correlated predictors, these elements represent an "inflation factor" for the variance of the regression coefficients, due to the redundancy of the predictors.

See also, multiple regression.

Variance, population variance The variance of a list is the square of the standard deviation of the list, that is, the average of the squares of the deviations of the numbers in the list from their mean. The variance of a random variable X, Var(X), is the expected value of the squared difference between the variable and its expected value: Var(X) = $E((X - E(X))^2)$. The variance of a random variable is the square of the standard error (SE) of the variable.

Venn diagram A pictorial way of showing the relations among sets or events. The universal set or outcome space is usually drawn as a rectangle; sets are regions within the rectangle. The overlap of the regions corresponds to the intersection of the sets. If the regions do not overlap, the sets are disjoint. The part of the rectangle included in one or more of the regions corresponds to the union of the sets.

Voronoi The Voronoi tessellation graph plots values of two variables X and Y in a scatterplot, then divides the space between individual data points into regions such that the boundaries surrounding each data point enclose an area that is closer to that data point than to any other neighboring points.

Voronoi labelling This algorithm assigns labels to units based upon the labels of the training cases that are "assigned" to that unit. Assigned cases are those that are nearer to this unit than to any other (i.e. those that would be classified by this unit if using the 1-NN classification scheme). These are the cases in the Voronoi neighborhood of the unit. The class of the majority of the training cases is used to label the unit, provided that at least a given minimum proportion of the training cases belong to this majority. If not, a blank label is applied, signifying an "unknown" class.

Voronoi scatterplot This specialized univariate scatterplot is more an analytic technique than just a method to graphically present data. The solutions it offers, help to model a variety of phenomena in natural and social sciences (e.g., Coombs, 1964; Ripley, 1981). The program divides the space between the individual data points represented by XY coordinates in 2D space. The division is such that each of the data points is surrounded by boundaries including only the area that is closer to its respective "center" data point than to any other data point.

The particular ways in which this method is used depends largely on specific research areas, however, in many of them, it is helpful to add additional dimensions to this plot by using categorization options.

See also, Data Reduction.

Voting See bagging.

W

Wald statistic The results Scrollsheet with the parameter estimates for the Cox proportional hazard regression model includes the so-called Wald statistic, and the p level for that statistic. This statistic is a test of significance of the regression coefficient; it is based on the asymptotic normality property of maximum likelihood estimates, and is computed as:

$$W = \beta^* 1/\mathrm{Var}(\beta) * \beta$$

In this formula, β stands for the parameter estimates, and Var(β) stands for the asymptotic variance of the parameter estimates. The Wald statistic is tested against the Chi-square distribution.

Way table This is a two-way table where the numbers of levels of the row- and column-classifications are each 2. If the row- and column- classifications each divide the observational units into subsets, then it is likely that it will be useful to analyse the data using the fisher test.

Weibull and reliability/failure time analysis In this context reliability is defined as the function that describes the probability of failure (or death) of an item as a function of time. Thus, the reliability function (commonly denoted as R(t)) is the complement to the cumulative distribution function (i.e., R(t)=1-F(t)); the reliability function is also sometimes referred to as the survivorship or survival function (since it describes the probability of not failing or surviving until a certain time t).

Weibull distribution The Weibull distribution has density function (for positive parameters b, c, and):

$$f(x) = c/b*[(x-\theta)/b]^{c-1} * e^{\wedge}\{-[(x-\theta)/b]^{c}\}$$

$$\theta < x,\ b > 0,\ c > 0$$

Where b is the scale parameter of the distribution

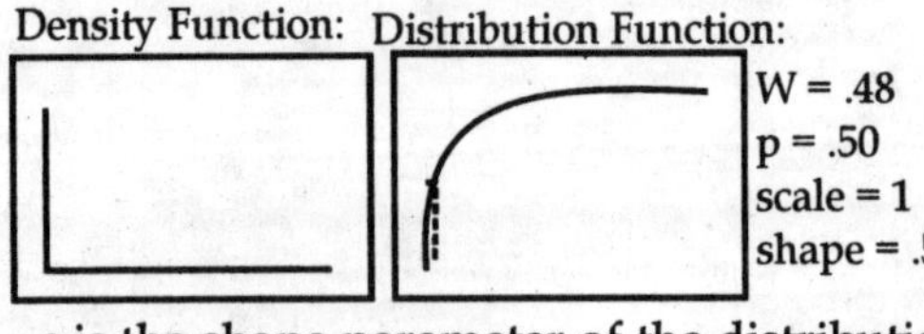

c is the shape parameter of the distribution

θ is the location parameter of the distribution

e is the base of the natural logarithm, sometimes called Euler's E.

Weigend weight regularization A common problem in neural network training (particularly of multilayer perceptrons) is over-fitting. A network with a large number of weights in comparison with the number of training cases available can achieve a low training error by modeling a function that fits the training data well despite failing to capture the underlying model. An over-fitted model typically has high curvature, as the function is contorted to pass through the points, modeling any noise in addition to the underlying data.

There are several approaches in neural network training to deal with the over-fitting problem (Bishop, 1995). These approaches are listed below.

Select a neural network with just enough units to model the underlying function. The problem with this approach is determining the correct number of units, which is problem-dependent.

Add some noise to the training cases during training (altering the noise on each case each epoch): this "blurs" the position of the training data, and forces the network, to model a smoothed version of the data.

Stop training (see Stopping Conditions) when the selection error begins to rise, even if the training error continues to fall. This event is a sure sign that the network is beginning to over-fit the data.

Use a regularization technique, which explicitly penalizes networks with large curvature, thus encouraging the development of a smoother model.

The last technique mentioned is regularization, and this section describes Weigend weight regularization (Weigend et. al., 1991).

A multilayer perceptron model with sigmoid (logistic or hyperbolic tangent) activation functions has higher curvature if the weights are larger. You can see this by considering the shape of the sigmoid curve: if you just look at a small part of the central section, around the value 0.0, it is "nearly linear," and so a network with very small weights will model a "nearly linear" function, which has low curvature. As an aside, note that during training the weights are first set to small values (corresponding to a low curvature function), and then (at least some of them) diverge. One way to promote low curvature therefore is to encourage smaller weights.

Weigend weight regularization does this by adding an extra term to the error function, which penalizes larger weights. Hence the network tends to develop the larger weights that it needs to model the problem, and the others are driven toward zero. The technique can be used with any multilayer perceptron training algorithms (back propagation, conjugate gradient descent, Quasi-Newton Method, quick propagation, and Delta-bar-Delta) apart from Levenberg-Marquardt, which makes its own assumptions about the error function.

The technique is commonly referred to as Weigend weight elimination, as it is possible, once weights become very small, to simply remove them from the network. This is an extremely useful technique for developing models with a "sensible" number of hidden units, and for selecting input variables.

Once a model has been trained with Weigend regularization and excess inputs and hidden units removed, it can be further trained with Weigend regularization turned off, to "sharpen up" the final solution.

Weigend regularization can also be very helpful in that it tends to prevent models from becoming over-fitted.

Note: When using Weigend regularization, the error on the progress graph includes the Weigend penalty factor. If you compare a network trained with Weigend to one without, you may get a false impression that the Weigend-trained network is under-performing. To compare such networks, view the error reported in the summary statistics on the model list (this does not include the Weigend error term).

Technical details The Weigend error penalty is given by:

$$\lambda \sum_i \frac{w_i^2}{w_o + w_i^2}$$

where ? is the Regularization coefficient, w_i is each of the weights, and w_o is the Scale coefficient.

The error penalty is added to the error calculated by the network's error function during training, and its derivative is added to the weight's derivative. However, the penalty is ignored when running a network.

The regularization coefficient is usually manipulated to adjust the selective pressure to prune units. The relationship between this coefficient and the number of active units is roughly logarithmic, so the coefficient is typically altered over a wide range (0.01-0.0001, say).

The scale coefficient defines what is a "large" value to the algorithm. The default setting of 1.0 is usually reasonable, and it is seldom altered.

A feature of the Weigend error penalty is that it does not just penalize larger weights. It also prefers to tolerate an uneven mix of some large and some small weights, as opposed to a number of medium-sized weights. It is this property that allows it to "eliminate" weights.

Weighted least squares (in regression) In some cases it is desirable to apply differential weights to the observations in a regression analysis, and to compute so-called weighted least squares regression estimates. This method is commonly applied when the variances of the residuals are not constant over the range of the independent variable values. In that case, one can apply the inverse values of the variances for the residuals as weights and compute weighted least squares estimates. (In practice, these variances are usually not known, however, they are often proportional to the values of the independent variable(s), and this proportionality can be exploited to compute appropriate case weights.) Neter, Wasserman, and Kutner (1985) describe an example of such an analysis.

Wilcoxon Mann Whitney test The Wilcoxon Mann-Whitney Test is one of the most powerful of the non-parametric tests for comparing two populations. It is used to test the null hypothesis that two populations have identical distribution functions against the alternative hypothesis that the two distribution functions differ only with respect to location (median), if at all.

The Wilcoxon Mann-Whitney test does not require the assumption that the differences between the two samples are normally distributed.

In many applications, the Wilcoxon Mann-Whitney Test is used in place of the two sample t-test when the normality assumption is questionable.

This test can also be applied when the observations in a sample of data are ranks, that is, ordinal data rather than direct measurements.

Wilcoxon signed ranks test The Wilcoxon Signed Ranks test is designed to test a hypothesis about the location (median) of a population distribution. It often involves the use of matched pairs, for example, before and after data, in which case it tests for a median difference of zero.

The Wilcoxon Signed Ranks test does not require the assumption that the population is normally distributed.

In many applications, this test is used in place of the one sample t-test when the normality assumption is questionable. It is a more powerful alternative to the sign test, but does assume that the population probability distribution is symmetric.

This test can also be applied when the observations in a sample of data are ranks, that is, ordinal data rather than direct measurements.

Wilcoxon test(1) Named after the statistician F, Wilcoxon, This test applies to an experimental design involving two repeated measure observations on a common set of experimental units, which need be only ordinal-scale. The purpose is to measure shift in scale location between the two levels of the repeated measure distinction. The test statistic is derived from the set of differences between the two levels of the repeated measure distinction - one difference score for each observational unit. The procedure is somewhat variable between authors, although the variants each correspond to valid well-sized exact test(1)s. Wilcoxon's original procedure commences by discarding entirely the observations from any experimental units for which the data values are equal at each level of the repeated measure comparison. Thus or otherwise, the next step is ranking the differences, providing a rank for each retained experimental unit; the ranks are according to the absolute values of the differences. The ranks are summed separately into two or three categories : negative differences; zero differences (if any); positive differences. The test statistic is the smaller of the outer categories, plus an adjustment for the middle (zero-difference) category. Also See pitman permutation test(2).

Wilcoxon test(2) Named after the statistician F, Wilcoxon, This is a test for an experimental design involving two independent groups of experimental units, where data need be only ordinal-scale. The purpose is to measure shift in scale location between the two groups. The test statistic is the sum, for a nominated group, of the ranks of the data for the groups combined. This test has an equivalent test statistic to that for the mann-whitney test, so the two tests must always agree. Also see : pitman permutation test(1).

Win frequencies In a Kohonen network, the number of times that each radial unit is the winner when the data set is executed. Units which win frequently represent cluster centers in the topological map.

Window The window factor is the number of epochs across which the error must fail to improve by the specified amount, before the algorithm is deemed to have slowed down too much and is stopped.

By default the window is zero, which means that the minimum improvement stopping condition is not used at all.

Wire A wire is a line, usually curved, used in a path diagram to represent variances and covariances of exogenous variables.

Working hypothesis A reformulation of a research hypothesis in terms of indicators rather than concepts

Wrong Number of cases erroneously assigned to another class.

X

X variable An independent variable in a regression analysis.

X11 output: A 1. original series This X-11 table shows the original series, prior to any initial user-defined or trading-day adjustment. Note that for quarterly series, no prior adjustment factors can be specified, and the original series will be shown as table B 1.

X11 output: A 2. prior monthly adjustment factors For X-11 monthly series, the user may specify a second series that contains prior monthly adjustment factors, for example, in order to adjust for an unusual holiday etc. The factors specified here will be subtracted from the original series for additive models, or will be used to divide the original series if multiplicative seasonal adjustment was requested (thus, the values in this series must be unequal to zero in that case).

X11 output: A 3. original series adjusted by prior monthly adjustment factors In this X-11 monthly series, the factors specified in A 2 will be subtracted from the original series (additive adjustment) or they will be used to divide the values in the original series (multiplicative adjustment). The resulting adjusted series is shown in this table.

X11 output: A 4. prior trading day adjustment factors This X-11 table is only available (applicable) when prior trading-day adjustment factors and a multiplicative model were specified. The user may specify a weight for each day (Monday through Friday); those weights are then proportionately adjusted so that they add to 7. The series (A 1 or A 3) is then divided by monthly calendar factors that are computed based on the number of the respective days in the respective month. Note that by default, the calendar factors are also adjusted for different lengths of different months; however, the length of month variability can also be included in the calendar factors (in which case a constant length of month of 30.4375 is used).

X11 output: B 1. prior adjusted series or original series This X-11 table shows the original series, or the initial adjusted series, depending on whether or not prior monthly adjustment factors and/or trading day adjustment factors where specified (for quarterly X-11, B 1 is always the original series).

X11 output: B 10. seasonal factors After replacing extreme values by the corresponding B 9 values, a 7-term weighted moving average is applied to the S-I differences (ratios) in B 8. The resulting estimate of the seasonal factors is then adjusted so that the sum for each year is equal to zero (additive model) or 12.0 (multiplicative model).

X11 output: B 11. seasonally adjusted series This X-11 table is the same as B 6, except that the seasonal factors in B 10 are used.

X11 output: B 13. irregular series The trend-cycle estimates in B 7 are subtracted from the seasonally adjusted series in B 11 (additive model), or the B 7 values are used to divide the series in B 11 (multiplicative model). The resulting series is an improved estimate of the irregular series.

X11 output: B 14. extreme irregular values excluded from trading day regression The months in the series are sorted into different groups, depending on the particular day when the month begins (30-day, 31-day months, and Februarys are treated separately). Then extreme values (beyond 2.5 * s; different s values can also be specified) are identified within each type of month in a two-step procedure. The final extreme values that will be excluded are shown in this X-11 table.

X11 output: B 15. preliminary trading day regression After removing the B 14 extreme values from B 13, least squares estimates for the seven daily weights are computed.

X11 output: B 16. trading day adjustment factors derived from regression coefficients From the trading-day regression weights, monthly adjustment factors are computed based on the number of particular trading days (i.e., Mondays, Tuesdays, etc.) in the respective months. These factors are printed in this X-11 table, and are then used to adjust (i.e., subtracted from or divided into) the B 13 irregular series for trading-day variation.

X11 output: B 17. preliminary weights for irregular component The estimates of the irregular component (in B 13 or adjusted by B 16, depending on whether or not a trading-day adjustment was performed) are further refined by computing graduated weights for extreme values, depending on their relative (in terms of a sliding 5-year s) distance from 0. Specifically, a process analogous to that described in B 4 above is used. This X-11 table (B17) contains the resulting adjustment factors.

X11 output: B 18. trading day factors derived from combined daily weights This X-11 table contains the final trading day adjustment factors, computed from the least squares trading-day weights in B 15 and/or the prior trading-day weights in A 4.

X11 output: B 19. original series adjusted for trading day and prior variation The values in B 18 are used to adjust the original (adjusted) series (in A 1, A 3, or B 1, depending on whether or not prior adjustment factors were specified). Specifically, the values in B 18 are subtracted from (additive model) or divided into (multiplicative model) the original series.

X11 output: B 3. unmodified S I differences or ratios An initial estimate of the combined irregular and seasonal component is obtained by subtracting B 2 from B 1 (additive model) or dividing B 1 by B 2 (multiplicative model).

X11 output: B 4. replacement values for extreme s i differences (ratios) First a preliminary estimate of the X-11 seasonal component is computed by applying a weighted 5-term moving average separately to the B 3 values for each month. Then a centered 12-term moving average of the preliminary factors for the entire series is computed, and the resulting values are adjusted to sum to zero (additive model) or 12.0 (multiplicative model) within each year. Next an initial estimate of the irregular component is obtained by subtracting from the S-I differences (additive model) or dividing the S-I ratios by the initial estimate for the seasonal component. For the resulting initial estimate of the irregular component, a 5-year sliding standard deviation (s, sigma) is computed, and extreme values in the central year that are beyond 2.5*s are removed. The 5-year sliding s is then recomputed and the process repeated; however, this time a zero weight is assigned to irregular values beyond 2.5*s, a full weight is assigned for values within 1.5*s, and linearly graduated weights between zero and one are assigned for values between 1.5 and 2.5 * s. Values receiving less than full weights are then recomputed as the average of the respective value times its weight and the nearest two full- weight values preceding and following the respective value in that month. Table B 4 shows the final replaced (re-computed) values, and the sliding 5-year s's.

X11 output: B 5. seasonal factors The extreme values in the B 3 series are replaced by the values shown in B 4. From this X-11 series, preliminary seasonal factors are derived by applying a 5-term moving average to each month separately; then a 12-term moving average is computed for the entire series, and the resulting values adjusted to sum to zero (additive model) or 12.0 (multiplicative model) within each year.

X11 output: B 6. seasonally adjusted series The preliminary seasonally adjusted series is obtained by subtracting from B 1 (additive model) or dividing B 1 (multiplicative model) by the seasonal factors in B 5.

X11 output: B 7. trend cycle The X 11 seasonally adjusted series (B 6) is smoothed via a variable moving average procedure. Optionally, extremes can be removed from the smoothed series by a process analogous to that described under B 4. In general, the so- called Henderson curve moving average is applied, which is a weighted moving average with the magnitudes of the weights following a bell-shaped curve. The choice of the appropriate length of the moving average is an important issue in the seasonal decomposition (i.e., the computation of the trend-cycle component). The general idea is to choose a longer moving average when there is a lot of random fluctuation in the data relative to the trend-cycle component, and to choose a shorter moving average when there is only relative little random fluctuation. By default the program will select a moving average transformation automatically. Specifically, first a preliminary 13-term Henderson (weighted) moving average of the seasonally adjusted series is computed (without extending to the ends of the series). A preliminary estimate of the irregular component is then computed by subtracting this series from (additive model) or dividing it into (multiplicative model) the seasonally adjusted series. Next, the average month- to-month difference (percent change) without regard to sign is computed for both the estimated irregular and trend-cycle components. The ratio of the average month-to-month differences (percent changes) in the two series reflects the relative importance of the irregular variations relative to the movements in the trend-cycle component. Depending on the value of this ratio, either a 9-term Henderson moving average is selected (if the ratio is between 0.0 and .99), a 13- term Henderson moving average is selected (if the ratio is between 1.0 and 3.49) or a 23-term Henderson moving average is selected (if the ratio is greater than 3.5).

X11 output: B 8. unmodified s i differences (ratios) This X-11 table is the same as B 3 except that it is based on the trend-cycle values computed in.

X11 output: B 9. replacement values for extreme S I differences (ratios) This X-11 table is the same as B 4 except that the differences (ratios) in B 8 are used to which a 7 term moving average is applied (to estimate the seasonal factors).

X11 output: C 1. original series modified by preliminary weights and adjusted for trading day and prior variation The series in B 19 (or B 1 if no trading-day adjustment was requested) is adjusted for extreme values by the weights computed in B 17 The resulting modified series in shown in this X-11 table (C 1).

X11 output: C 10. seasonal factors The seasonal factors are computed analogously to B 10, but based on the C 9 S-I differences (ratios).

X11 output: C 11. seasonally adjusted series The refined seasonally adjusted series is computed by subtracting from B 1 (additive model) or dividing B 1 by (multiplicative model) the values in C 10.

X11 output: C 13. irregular series The refined estimate of the irregular (random) component is computed by subtracting from C 11 (additive model) or dividing C 11 by (multiplicative model) the values in C 7.

X11 output: C 14. extreme irregular values excluded from trading day regression This table is analogous to table B 14, and it shows the extreme irregular values (usually beyond 2.5 * s) after re-applying the trading-day routine (based on the monthly trading-day factors shown in B 16).

X11 output: C 15. final trading day regression This X-11 table is the same as B 15, except that the computations are based on the values from table C 13.

X11 output: C 16. final trading day adjustment factors derived from regression X11 output: coefficients This X-11 table is analogous to B 16, except that the factors are subtracted from (additive case) or divided into (multiplicative case) the values from table C 13.

X11 output: C 17. final weights for irregular component This table is analogous to table B 17, except that it is computed based on the values in C 16 (or C 13 if no trading day adjustment is requested).

X11 output: C 18. final trading day factors derived from combined daily weights This X-11 table is analogous to B 18, except that the final weights shown in C 15 are used in the computations.

X11 output: C 19. original series adjusted for trading day and prior variation The values in C 18 are used to adjust the original (adjusted) series (in A 3 or B 1). Specifically, the values in C 18 are subtracted from (additive model) or divided into (multiplicative model) the original series.

X11 output: C 2. trend cycle An estimate of the combined trend-cycle component is computed from C 1 by applying a centered 12-term moving average.

X11 output: C 4. modified SI differences (ratios) To obtain the refined S-I differences (ratios), the values in C 2 are subtracted from (additive model) or divided into (multiplicative model) the modified series in C 1.

X11 output: C 5. seasonal factors These values are the same as those in B 5, except that the C 4 differences (ratios) are used.

X11 output: C 6. seasonally adjusted series The preliminary seasonally adjusted series is computed by subtracting C 5 from (or dividing C 5 into) C 1.

X11 output: C 7. trend cycle The seasonally adjusted series (C 6) is smoothed via a variable moving average procedure (the same procedure used for B 7, see also Shiskin, Young, & Musgrave, 1967, for details) to derive the preliminary estimate of the trend-cycle component.

X11 output: C 9. modified s i differences (ratios) The modified S-I differences (ratios) are computed by subtracting C 7 from (additive models) or dividing C 7 into (multiplicative models) the C 1 series.

X11 output: D 1. original series modified by final weights and adjusted for trading day

and prior variation This X-11 table is analogous to C 1, except that the C 17 weights and C 19 adjusted series are used in the computations.

X11 output: D 10. final seasonal factors This X-11 table is computed analogously to the values in B 10, except that it is computed based on the values reported in D 8 and D 9.

X11 output: D 11. final seasonally adjusted series The final seasonally adjusted series is computed by subtracting D 10 from C 19 (additive model) or dividing C 19 by D 10 (multiplicative model).

X11 output: D 12. final trend cycle These values are computed by subtracting D 10 from D 1 (additive model), or by dividing D 1 by D 10 (multiplicative model). X11 output: D 13. final irregular These values are computed by subtracting D 12 from D 11 (additive model), or by dividing D 11 by D 12 (multiplicative model). X11 output: D 2. trend cycle A 12-term moving average of D 1 is computed to estimate the trend-cycle component.

X11 output: D 4. modified S I differences (ratios) The modified S-I differences (ratios) are computed by subtracting D 2 from (additive model) or dividing D 2 into (multiplicative model) the values in D 1.

X11 output: D 5. seasonal factors This X-11 table is computed analogously to B 5, except that the computations are based on the values in D 4.

X11 output: D 6. seasonally adjusted series The values in this table are computed by subtracting D 5 from D 1 (additive model) or dividing D 1 by D5 (multiplicative model).

X11 output: D 7. trend cycle The values in this X-11 table are computed analogously to those in B 7, except that the computations are based on the values in D 6.

X11 output: D 8. final unmodified s i differences (ratios) The values in the D 7 series are subtracted from (additive model) or divided into (multiplicative case) the values in C 19 (or B 1 if no adjustment for trading-day variation is applied). Then an analysis of variance by month (or quarter) is performed on this series, in order to test for the presence of stable significant seasonality.

X11 output: D 9. final replacement values for extreme S I differences (ratios) The values in D 7 are subtracted from (additive model) or divided into (multiplicative model) D 1; values that are not identical to the corresponding entries in D 8 are then reported. Also, for each month, the year-to-year difference (additive model) or percent change (multiplicative mode) in the estimates of the irregular and the seasonal components and their ratio (called MSR, moving seasonality ratio) are computed. The MSR may be useful in order to determine the amount of moving seasonality present in each month.

X11 output: E 1. modified original series The values in this X-11 table are computed by replacing in the original series extreme values (identified by a zero weight in C 17) by the values predicted from the final trend-cycle, seasonal, trading-day (if applicable), and prior adjustment (if applicable) components.

X11 output: E 2. modified seasonally adjusted series These values are computed by replacing in the final seasonally adjusted series (D 11) extreme values (identified by a zero weight in C 17) with the D 12 final trend-cycle values.

X11 output: E 3. modified irregular series The values in this X-11 table are computed by replacing the values in D 13 with zero (additive model) or 1.0 (multiplicative model) if they were identified as extremes (i.e., assigned zero weight) in C 17.

X11 output: E 4. differences (ratios) of annual totals These values are computed as the differences (additive model) or ratios (multiplicative model) of the annual totals of (a) the original series B 1 and the final seasonally adjusted series D 11, (b) the modified original series E 1 and the modified seasonally adjusted series E 2.

X11 output: E 5. differences (percent changes) in original series The values in this X-11 table are computed as the month-to-month (quarter-to-quarter) differences (additive model) or percent changes (multiplicative model) in B 1. **X11 output: E 6. differences (percent changes) in final seasonally adjusted series** These values are the month-to-month (quarter-to-quarter) differences (additive model) or percent changes (multiplicative model) in D 11.

X11 output: F 1. MCD (QCD) moving average The values in this series are computed by applying an unweighted moving average to the final seasonally adjusted series (D 11). The width of the smoothing window is determined by the month (quarter) for cyclical dominance, or MCD (QCD) for short. The MCD (QCD) is computed as the average span at which the changes in the random component are equal to the changes in the trend-cycle component

X11 output: F 2. summary measures Several final summary X-11 tables are computed:

The average differences (additive model) or percent changes (multiplicative model) are computed without regard to sign across spans 1, 2, 3 ..., 12 months (or four quarters) for the following series: Original series A 1 (B 1), final seasonally adjusted series (D 11), final irregular series (D 13), final trend-cycle (D 12), final seasonal factors (D 10), final prior monthly adjustment factors (A 2, monthly X-11 only), final trading-day adjustment factors (C 18, monthly X-11 only), modified original series (E 1), modified seasonally adjusted series (E 2), modified irregular series (E 3).

Next a table of relative contributions of the different components to the differences (additive model) or percent changes (multiplicative model) in the original series are computed.

The next table reports the average duration of run (the average number of consecutive monthly changes in the same direction; "no change" is counted as a change in the same direction) for the following series: Final seasonally adjusted series (D 11), final irregular series (D 13), final trend-cycle (D 12), and the MCD (QCD) moving average (F 1).

Finally, the means and standard deviations of differences (additive model) or percent changes (multiplicative model) are computed across different spans for each of the series mentioned above.

X11 output: G 1. chart This line graph will show the final seasonally adjusted series and final trend-cycle components (D 11 and D 12, respectively).

X11 output: G 2. chart This line graph will show the final S-I differences (additive model) or ratios (multiplicative model) with the extremes, the final S-I differences (ratios) without extremes, and the final seasonal factors (i.e., D 8, D 9, and D 10, respectively), categorized by month (X-11 monthly) or quarter (X-11 quarterly).

X11 output: G 3. chart This plot shows the same values as G 2; however, this line plot shows those values in chronological order.

X11 output: G 4. chart This is a line graph of the final irregular and final modified irregular series (D 13 and E 3, respectively).

XML (extensible markup language) XML (short for Extensible Markup Language) is a specification language developed by the World Wide Web Consortium (W3C). XML is a language standard designed especially for Web documents, to allow programmers to create their own customized tags, thus enabling the definition, transmission, validation, and interpretation of data between applications and between organizations. A special version of XML is PMML.

Xor exclusive disjunction Xor is an operation on two logical propositions. If p and q are two propositions, (p xor q) is a proposition that is true if either p is true or if q is true, but not both. (p xor q) is logically equivalent to ((p or q) and not (p and q)).

Y

Y variable The dependent variable in a regression analysis

Yates corrected chi square The approximation of the Chi-square statistic in small 2 x 2 tables can be improved by reducing the absolute value of differences between expected and observed frequencies by 0.5 before squaring (Yates' correction). This correction, which makes the estimation more conservative, is usually applied when the table contains only small observed frequencies, so that some expected frequencies become less than 10.

Yield The number of non-defects can be considered the Yield of the process. Six Sigma calculators will compute the number of defects per million opportunities (DPMO) as well as the yield, expressed as the percent of the area under the normal curve that falls below the upper specification limit.

Z

Z distribution (standard normal) The Z distribution (or standard normal distribution) function is determined by the following formula:

f(x) = 1/[(2p)1/2] * e**{-1/2*x2}

-8 < x < 8

where

e is the base of the natural logarithm, sometimes called Euler's e

pi is the constant Pi (3.14...)

Note that this distribution is simply a normal distribution where the mean is zero and the standard deviation is one.

The Z distribution is commonly used in hypothesis testing for large samples or in situations where the standard deviation is known.

Z score Standardisation of a value by subtracting the variable's mean from the value and dividing by the variable's standard deviation. A distribution of z scores always has a mean of zero and a standard deviation of unity.

Z statistic A Z statistic is a test statistic whose distribution under the null hypothesis has expected value zero and can be approximated well by the normal curve. Usually, Z statistics are constructed by standardizing some other statistic. The Z statistic is related to the original statistic by

Z = (original - expected value of original)/ SE(original).

Z test An hypothesis test based on approximating the probability histogram of the Z statistic under the null hypothesis by the normal curve.